WAVES, WIND
AND WEATHER

WAVES, WIND AND WEATHER

Nathaniel Bowditch

Selected from

AMERICAN PRACTICAL NAVIGATOR

David McKay Company, Inc.

New York

WAVES, WIND AND WEATHER

Selected from
American Practical Navigator

Published by
David McKay Company, Inc.
750 Third Avenue
New York, New York 10017

No portion of this edition
may be mechanically reproduced by any means
without approval in writing from
the publisher.

1977

Library of Congress Cataloging in Publication Data

Bowditch, Nathaniel, 1773–1838.
  Waves, wind and weather.
  1. Meteorology, Maritime. 2. Weather. 3. Ocean-
ography. I. Title.
QC994.B66 1977   623.89   77–8040
ISBN 0–679–50753–1

MANUFACTURED IN THE UNITED STATES OF AMERICA
Design by R. R. Duchi

Contents

THE SEA 1
 Ocean Waves 3
 The Oceans 19
 Surf 33
 Ocean Currents 38
 Tides and Tidal Currents 50
 Ice in the Sea 63

THE SKY 85
 Weather and Weather Forecasts 87
 Weather Observations 121
 Clouds 145
 Tropical Cyclones 149

Appendix A—*Glossary* 169
Appendix B—*Weather Broadcasts* 183
Appendix C—*Beaufort Scale* 186

An insert on "International Cloud Code Guide for the Mariner" following page 146

The Sea

=1=

Ocean Waves

Introduction—Undulations of the surface of the water, called **waves,** are perhaps the most widely observed phenomenon at sea, and possibly the least understood by the average seaman. The mariner equipped with a knowledge of the basic facts concerning waves is able to use them to his advantage, and either avoid hazardous conditions or operate with a minimum of danger if such conditions cannot be avoided.

Causes of Waves—Waves on the surface of the sea are caused principally by wind, but other factors, such as submarine earthquakes, volcanic eruptions, and the tide, also cause waves. If a breeze of less than two knots starts to blow across smooth water, small wavelets called **ripples** form almost instantaneously. When the breeze dies, the ripples disappear as suddenly as they formed, the level surface being restored by surface tension of the water. If the wind speed exceeds two knots, more stable **gravity waves** gradually form, and progress with the wind.

While the generating wind blows, the resulting waves may be referred to as **sea.** When the wind stops or changes direction, the waves that continue on without relation to local winds are called **swell.**

Unlike wind and current, waves are not deflected appreciably by the rotation of the earth, but move in the direction in which the generating wind blows. When this wind ceases, friction and spreading cause the waves to be reduced in height, or **attenuated,** as they move across the surface. However, the reduction takes place so slowly that swell continues until it reaches some obstruction, such as a shore.

When sufficient data on wind conditions are available, the swell and state of the sea a day or more in advance can be predicted. Such forecasts have been found useful in wartime offshore unloading operations. The U.S. Navy Hydrographic Office forecasts sea swell conditions.

Wave Characteristics—Ocean waves are very nearly in the shape of an inverted **cycloid,** the figure formed by a point inside the rim of a wheel rolling along a level surface. This shape is shown in figure 1–1.

Figure 1–1 *A typical sea wave.*

The highest parts of waves are called **crests,** and the intervening lowest parts, **troughs.** Since the crests are steeper and narrower than the troughs, the mean or still water level is a little lower than halfway between the crests and troughs. The vertical distance between trough and crest is called **wave height,** labeled H. The horizontal distance between successive crests, measured in the direction of travel, is called **wave length,** labeled L. The time interval between passage of successive crests at a stationary point is called **wave period (P).** Wave height, length, and period depend upon a number of factors, such as the wind speed, the length of time it has blown, and its **fetch** (the straight distance it has traveled over the surface). Table 1–1 indicates the relationship between wind speed, fetch, length of time the wind blows, wave height, and wave period in deep water.

If the water is deeper than one-half the wave length (L), this length in feet is theoretically related to period (P) in seconds by the formula

$$L = 5.12P^2.$$

The actual value has been found to be a little less than this for swell, and about two-thirds the length determined by this formula for sea. When the waves leave the generating area and continue as free waves, the wave length and period continue to increase, while the height decreases. The rate of change gradually decreases.

The speed (S) of a free wave in deep water is nearly independent of its height or steepness. For swell, its relationship in knots to the period (P) in seconds is given by the formula

$$S = 3.03P.$$

The relationship for sea is not known.

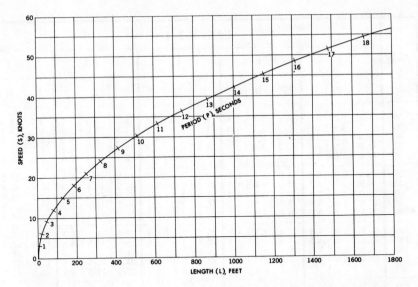

Figure 1-2 *Relationship between speed, length, and period of waves in deep water, based upon the theoretical relationship between period and length.*

The theoretical relationship between speed, wave length, and period is shown in figure 1-2. Thus, as waves continue on beyond the the generating area, the period, length, and speed all increase, providing some indication of the distance of the generating area. However, the time needed for a wave system to travel some distance is *double* that which would be indicated by the speed of individual waves. This is because the front wave gradually disappears and transfers its energy to succeeding waves. The process is followed by each front wave in succession, at such a rate that the wave *system* advances at a speed which is just *half* that of *individual* waves. This process can be seen in the bow wave of a vessel. The speed at which the wave system advances is called **group velocity.**

Because of the existence of many independent wave systems at the same time, the sea surface acquires a complex and irregular pattern. Also, since the longer waves outrun the shorter ones, the resulting interference adds to the complexity of the pattern. The process of interference, illustrated in figure 1-3, is duplicated many times in the sea, being the principal reason that successive waves are not of

TABLE 1–1

Fetch	3			4			5			6			7	
	T	H	P	T	H	P	T	H	P	T	H	P	T	H
10	4.4	1.8	2.1	3.7	2.6	2.4	3.2	3.5	2.8	2.7	5.0	3.1	2.5	6.0
20	7.1	2.0	2.5	6.2	3.2	2.9	5.4	4.9	3.3	4.7	7.0	3.8	4.2	8.6
30	9.8	2.0	2.8	8.3	3.8	3.3	7.2	5.8	3.7	6.2	8.0	4.2	5.8	10.0
40	12.0	2.0	3.0	10.3	3.9	3.6	8.9	6.2	4.1	7.8	9.0	4.6	7.1	11.2
50	14.0	2.0	3.2	12.4	4.0	3.8	11.0	6.5	4.4	9.1	9.8	4.8	8.4	12.2
60	16.0	2.0	3.5	14.0	4.0	4.0	12.0	6.8	4.6	10.2	10.3	5.1	9.6	13.2
70	18.0	2.0	3.7	15.8	4.0	4.1	13.5	7.0	4.8	11.9	10.8	5.4	10.5	13.9
80	20.0	2.0	3.8	17.0	4.0	4.2	15.0	7.2	4.9	13.0	11.0	5.6	12.0	14.5
90	23.6	2.0	3.9	18.8	4.0	4.3	16.5	7.3	5.1	14.1	11.2	5.8	13.0	15.0
100	27.1	2.0	4.0	20.0	4.0	4.4	17.5	7.3	5.3	15.1	11.4	6.0	14.0	15.5
120	31.1	2.0	4.2	22.4	4.1	4.7	20.0	7.8	5.4	17.0	11.7	6.2	15.9	16.0
140	36.6	2.0	4.5	25.8	4.2	4.9	22.5	7.9	5.8	19.1	11.9	6.4	17.6	16.2
160	43.2	2.0	4.9	28.4	4.2	5.2	24.3	7.9	6.0	21.1	12.0	6.6	19.5	16.5
180	50.0	2.0	4.9	30.9	4.3	5.4	27.0	8.0	6.2	23.1	12.1	6.8	21.3	17.0
200				33.5	4.3	5.6	29.0	8.0	6.4	25.4	12.2	7.1	23.1	17.5
220				36.5	4.4	5.8	31.1	8.0	6.6	27.2	12.3	7.2	25.0	17.9
240				39.2	4.4	5.9	33.1	8.0	6.8	29.0	12.4	7.3	26.8	17.9
260				41.9	4.4	6.0	34.9	8.0	6.9	30.5	12.6	7.5	28.0	18.0
280				44.5	4.4	6.2	36.8	8.0	7.0	32.4	12.9	7.8	29.5	18.0
300				47.0	4.4	6.3	38.5	8.0	7.1	34.1	13.1	8.0	31.5	18.0
320							40.5	8.0	7.2	36.0	13.3	8.2	33.0	18.0
340							42.4	8.0	7.3	37.6	13.4	8.3	34.2	18.0
360							44.2	8.0	7.4	38.8	13.4	8.4	35.7	18.1
380							46.1	8.0	7.5	40.2	13.5	8.5	37.1	18.2
400							48.0	8.0	7.7	42.2	13.5	8.6	38.8	18.4
420							50.0	8.0	7.8	43.5	13.6	8.7	40.0	18.7
440							52.0	8.0	7.9	44.7	13.7	8.8	41.3	18.8
460							54.0	8.0	8.0	46.2	13.7	8.9	42.8	19.0
480							56.0	8.0	8.1	47.8	13.7	9.0	44.0	19.0
500							58.0	8.0	8.2	49.2	13.8	9.1	45.5	19.1
550										53.0	13.8	9.3	48.5	19.5
600										56.3	13.8	9.5	51.8	19.7
650													55.0	19.8
700													58.5	19.8
750														
800														
850														
900														
950														
1000														

*BEAUFORT NUMBER

	8			9			10			11			Fetch
P	T	H	P	T	H	P	T	H	P	T	H	P	
3.4	2.3	7.3	3.9	2.0	8.0	4.1	1.9	10.0	4.2	1.8	10.0	5.0	10
4.3	3.9	10.0	4.4	3.5	12.0	5.0	3.2	14.0	5.2	3.0	16.0	5.9	20
4.6	5.2	12.1	5.0	4.7	15.8	5.5	4.4	18.0	6.0	4.1	19.8	6.3	30
4.9	6.5	14.0	5.4	5.8	17.7	5.9	5.4	21.0	6.3	5.1	22.5	6.7	40
5.2	7.7	15.7	5.6	6.9	19.8	6.3	6.4	23.0	6.7	6.1	25.0	7.1	50
5.5	8.7	17.0	6.0	8.0	21.0	6.5	7.4	25.0	7.0	7.0	27.5	7.5	60
5.7	9.9	18.0	6.4	9.0	22.5	6.8	8.3	26.5	7.3	7.8	29.5	7.7	70
6.0	11.0	18.9	6.6	10.0	24.0	7.1	9.3	28.0	7.7	8.6	31.5	7.9	80
6.3	12.0	20.0	6.7	11.0	25.0	7.2	10.2	30.0	7.9	9.5	34.0	8.2	90
6.5	12.8	20.5	6.9	11.9	26.5	7.6	11.0	32.0	8.1	10.3	35.0	8.5	100
6.7	14.5	21.5	7.3	13.1	27.5	7.9	12.3	33.5	8.4	11.5	37.5	8.8	120
7.0	16.0	22.0	7.6	14.8	29.0	8.3	13.9	35.5	8.8	13.0	40.0	9.2	140
7.3	18.0	23.0	8.0	16.4	30.5	8.7	15.1	37.0	9.1	14.5	42.5	9.6	160
7.5	19.9	23.5	8.3	18.0	31.5	9.0	16.5	38.5	9.5	16.0	44.5	10.0	180
7.7	21.5	23.5	8.5	19.3	32.5	9.2	18.1	40.0	9.8	17.1	46.0	10.3	200
8.0	22.9	24.0	8.8	20.9	34.0	9.6	19.1	41.5	10.1	18.2	47.5	10.6	220
8.2	24.4	24.5	9.0	22.0	34.5	9.8	20.5	43.0	10.3	19.5	49.0	10.8	240
8.4	26.0	25.0	9.2	23.5	34.5	10.0	21.8	44.0	10.6	20.9	50.5	11.1	260
8.5	27.7	25.0	9.4	25.0	35.0	10.2	23.0	45.0	10.9	22.0	51.5	11.3	280
8.7	29.0	25.0	9.5	26.3	35.0	10.4	24.3	45.0	11.1	23.2	53.0	11.6	300
8.9	30.2	25.0	9.6	27.6	35.5	10.6	25.5	45.5	11.2	24.5	54.0	11.8	320
9.0	31.6	25.0	9.8	29.0	36.0	10.8	26.7	46.0	11.4	25.5	55.0	12.0	340
9.1	33.0	25.0	9.9	30.0	36.5	10.9	27.7	46.5	11.6	26.6	55.0	12.2	360
9.3	34.2	25.5	10.0	31.3	37.0	11.1	29.1	47.0	11.8	27.7	55.5	12.4	380
9.5	35.6	26.0	10.2	32.5	37.0	11.2	30.2	47.5	12.0	28.9	56.0	12.6	400
9.6	36.9	26.5	10.3	33.7	37.5	11.4	31.5	47.5	12.2	29.6	56.5	12.7	420
9.7	38.1	27.0	10.4	34.8	37.5	11.5	32.5	48.0	12.3	30.9	57.0	12.9	440
9.8	39.5	27.5	10.6	36.0	37.5	11.7	33.5	48.5	12.5	31.8	57.5	13.1	460
9.9	41.0	27.5	10.8	37.0	37.5	11.8	34.5	49.0	12.6	32.7	57.5	13.2	480
10.1	42.1	27.5	10.9	38.3	38.0	11.9	35.5	49.0	12.7	33.9	58.0	13.4	500
10.3	44.9	27.5	11.1	41.0	38.5	12.2	38.2	50.0	13.0	36.5	59.0	13.7	550
10.5	47.7	27.5	11.3	43.6	39.0	12.5	40.3	50.0	13.3	38.7	60.0	14.0	600
10.7	50.3	27.5	11.6	46.4	39.5	12.8	43.0	50.0	13.7	41.0	60.0	14.2	650
11.0	53.2	27.5	11.8	49.0	40.0	13.1	45.4	50.5	14.0	43.5	60.5	14.5	700
	56.2	27.5	12.1	51.0	40.0	13.3	48.0	51.0	14.2	45.8	61.0	14.8	750
	59.2	27.5	12.3	53.8	40.0	13.5	50.6	51.5	14.5	47.8	61.5	15.0	800
				56.2	40.0	13.8	52.5	52.0	14.6	50.0	62.0	15.2	850
				58.2	40.0	14.0	54.6	52.0	14.9	52.0	62.5	15.5	900
							57.2	52.0	15.1	54.0	63.0	15.7	950
							59.3	52.0	15.3	56.3	63.0	16.0	1000

*Minimum Time (T) in hours that wind must blow to form waves of H significant height (in feet) and P period (in seconds). Fetch in nautical miles. Based upon the relationships given in H.O. Pub. No. 604, *Techniques for Forecasting Wind Waves and Swell*. See also H.O. Pub. No. 603, *Observing and Forecasting Ocean Waves*.

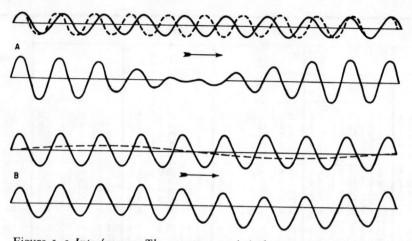

Figure 1–3 *Interference. The upper part of A shows two waves of equal height and nearly equal length traveling in the same direction. The lower part of A shows the resulting wave pattern. In B similar information is shown for short waves and long swell.*

the same height. The irregularity of the surface may be further accentuated by the presence of wave systems crossing at an angle to each other, producing peak-like rises.

In reporting average wave heights, the mariner has a tendency to neglect the lower ones. It has been found that the reported value is about the average for the highest one-third. This is sometimes called the "significant" wave height. The approximate relationship between this height and others, is as follows:

Wave	Relative height
Average	0.64
Significant	1.00
Highest 10 percent	1.29
Highest	1.87

Path of Water Particles in a Wave—As shown in figure 1–4, a particle of water on the surface of the ocean follows a somewhat circular orbit as a wave passes, but moves very little in the direction of motion of the wave. The common wave producing this action is called an **oscillatory wave.** As the crest passes, the particle moves forward, giving water the appearance of moving with the wave. As the trough passes, the motion is in the opposite direction. The radius

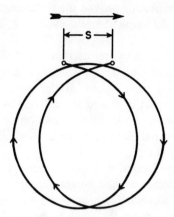

Figure 1–4 *Orbital motion and displacement, s, of a particle on the surface of deep water during two wave periods.*

of the circular orbit decreases with depth, approaching zero at a depth equal to about half the wave length. In shallower water the orbits become more elliptical, and in very shallow water, as at a beach, the vertical motion disappears almost completely.

Since the speed is greater at the top of the orbit than at the bottom, the particle is not at exactly its original point following passage of a wave, but has moved slightly in the direction of motion of the wave. However, since this advance is small in relation to the vertical displacement, a floating object is raised and lowered by passage of a wave, but moved little from its original position. If this were not so, a slow moving vessel might experience considerable difficulty in making way against a wave train. In figure 4–1, the forward displacement is greatly exaggerated.

Effects of Currents on Waves—A following current increases wave lengths and decreases wave heights. An opposing current has the opposite effect, decreasing the length and increasing the height. A strong opposing current may cause the waves to break. The extent of wave alteration is dependent upon the ratio of the still-water wave speed to the speed of the current.

Moderate ocean currents running at oblique angles to wave directions appear to have little effect, but strong tidal currents perpendicular to a system of waves have been observed to completely destroy them in a short period of time.

The Effect of Ice on Waves—When ice crystals form in sea water, internal friction is greatly increased. This results in smoothing of the sea surface. The effect of pack ice is even more pronounced. A vessel following a lead through such ice may be in smooth water even when a gale is blowing and heavy seas are beating against the outer edge of the pack. Hail is also effective in flattening the sea, even in a high wind.

Waves and Shallow Water—When a wave encounters shallow water, the movement of the individual particles of water is restricted by the bottom, resulting in reduced wave speed. If the wave approaches the shoal at an angle, each part is slowed successively as the depth decreases. This causes a change in direction of motion or **refraction,** the wave tending to become parallel to the depth curves. The effect is similar to the refraction of light.

As each wave slows, the next wave behind it, in deeper water, tends to catch up. As the wave length decreases, the height generally becomes greater. The lower part of a wave, being nearest the bottom, is slowed more than the top. This may cause the wave to become unstable, the faster-moving top falling or **breaking.** Such a wave is called a **breaker,** and a series of breakers, **surf.**

Swell passing over a shoal but not breaking undergoes a decrease in wave length and speed, and an increase in height. Such **ground swell** may cause heavy rolling if it is on the beam and its period is the same as the period of roll of a vessel, even though the sea may appear relatively calm. Figure 1–5 shows the approximate alteration of the characteristics of waves as they cross a shoal.

Energy of Waves—The potential energy of a wave is related to the vertical distance of each particle from its still-water position, and therefore moves with the wave. In contrast, the kinetic energy of a wave is related to the speed of the particles, being distributed evenly along the entire wave.

The amount of kinetic energy in even a moderate wave is tremendous. A four-foot, ten-second wave striking a coast expends more than 35,000 horsepower per mile of beach. For each 56 miles of coast, the energy expended equals the power generated at Hoover Dam. An increase in temperature of the water in the relatively narrow **surf zone** in which this energy is expended would seem to be indicated, but no pronounced increase has been measured. Apparently, any heat that may be generated is dissipated to the deeper water beyond the surf zone.

Wave Measurement Aboard Ship—With suitable equipment and adequate training, one can make reasonably reliable measurements of the height, length, period, and speed of waves. However, the mariner's estimates of height and length usually contain relatively large

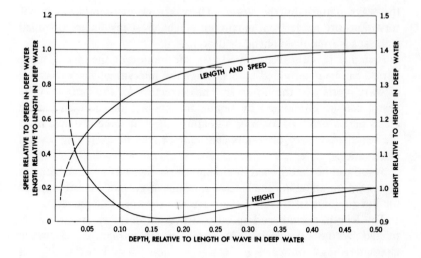

Figure 1–5 *Alteration of the characteristics of waves as they cross a shoal.*

errors. There is a tendency to underestimate the heights of low waves, and overestimate the heights of high ones. There are numerous accounts of waves 75 to 80 feet high, or even higher, although waves more than 55 feet high are very rare. Wave length is usually underestimated. The motions of the vessel from which measurements are made perhaps contribute to such errors.

Height. Measurement of wave height is particularly difficult. A microbarograph can be used if the wave is long enough to permit the vessel to ride up and down with it. If the waves are approaching from dead ahead or dead astern, this requires a wave length at least twice the length of the vessel. For most accurate results the instrument should be placed at the center of roll and pitch, to minimize the effects of these motions. Wave height can often be estimated with reasonable accuracy by comparing it with freeboard of the vessel. This is less accurate as wave height and vessel motion increase. If a point of observation can be found at which the top of a wave is in line with the horizon when the observer is in the trough, the wave height is equal to height of eye. However, if the vessel is rolling or pitching, this height at the moment of observation may be difficult to determine.

Length. The dimensions of the vessel can be used to determine wave length. Errors are introduced by perspective and disturbance of the wave pattern by the vessel. These errors are minimized if observations are made from maximum height. Best results are obtained if the sea is from dead ahead or dead astern.

Period. If allowance is made for the motion of the vessel, wave period can be determined by measuring the interval between passages of wave crests past the observer. The correction for the motion of the vessel can be eliminated by timing the passage of successive wave crests past a patch of foam or a floating object at some distance from the vessel. Accuracy of results can be improved by averaging several observations.

Speed can be determined by timing the passage of the wave between measured points along the side of the ship, if corrections are applied for the direction of travel of the wave and the speed of the ship.

The length, period, and speed of waves in deep waters are interrelated by the relationships indicated in the paragraph on "wave characteristics," on page 4. However, these should be used as a general guide only, because exact mathematical relationships have not been established. In the case of speed and period, there is evidence to indicate that for sea the relationship may be more nearly expressed by the formula $L = SP$ than by that given on page 00, although there is considerable doubt as to the exact relationship. There is no definite mathematical relationship between wave height and length, period, or speed.

Tsunamis are ocean waves produced by sudden, large-scale motion of a portion of the ocean floor or the shore, as by volcanic eruption, earthquake (sometimes called **seaquake** if it occurs at sea), or landslide. If they are caused by a submarine earthquake, they are usually called **seismic sea waves.** The point directly above the disturbance, at which the waves originate, is called the **epicenter.** Either a tsunami or a storm wave that overflows the land is popularly called a **tidal wave,** although it bears no relation to the tide.

If a volcanic eruption occurs below the surface of the sea, the escaping gases cause a quantity of water to be pushed upward in the shape of a dome or mound. The same effect is caused by the sudden rising of a portion of the bottom. As this water settles back, it creates a wave which travels at high speed across the surface of the ocean.

Tsunamis usually occur in series, gradually increasing in height until a maximum is reached between about the third and eighth wave. Following the maximum, they again become smaller. Waves may continue to form for several hours, or even for days.

In deep water the wave height of a tsunami is probably never greater than two or three feet. Since the wave length is usually considerably more than 100 miles, the wave is not conspicuous at sea. In the Pacific, where most tsunamis occur, the wave period varies between about 15 and 60 *minutes*, and the speed in deep water is more than 400 knots. The approximate speed can be computed by the formula

$$S = 0.6 \sqrt{gd} = 3.4 \sqrt{d},$$

where S is the speed in knots, g is the acceleration due to gravity (32.2 feet per second per second), and d is the depth of water in feet. This formula is applicable to any wave in water having a depth of less than half the wave length. For most ocean waves it applies only in shallow water, because of the relatively short wave length.

When a tsunami enters shoal water, it undergoes the same changes as other waves. The formula indicates that speed is proportional to depth of water. Because of the great speed of a tsunami when it is in relatively deep water, the slowing is relatively much greater than that of an ordinary wave crested by wind. Therefore, the increase in height is also much greater. Tsunamis 50 feet in height or higher have reached the shore, inflicting widespread damage. On April 1, 1946, seismic sea waves originating at an epicenter near the Aleutians spread over the entire Pacific. Scotch Cap Light on Unimak Island, 57 feet above sea level, was completely destroyed. Traveling at an average speed of 490 miles per hour, the waves reached the Hawaiian Islands in four hours and 34 minutes, where they arrived as waves 50 feet above the high water level, and flooded a strip of coast more than 1,000 feet wide at some places. They left a death toll of 173, and property damage of $25,000,000. Less destructive waves reached the shores of North and South America, and Australia, 6,700 miles from the epicenter.

After this disaster, a tsunami warning system was set up in the Pacific, even though destructive waves are relatively rare (averaging about one in 20 years in the Hawaiian Islands). The system consists of three sections. First, a number of seismograph stations to provide information for establishing the time and epicenter of quakes. Second, a group of tide stations to report any evidence of a tsunami. These stations are alerted when a quake is recorded at the seismograph stations. Third, a communication system which gives tsunami warnings high priority because of their speed and possible destructiveness. A travel time chart centered upon the Hawaiian Islands is used to to estimate time of arrival of the waves.

Fortunately, relatively few earthquakes produce seismic sea waves. The size of the waves that do form depends upon the nature and intensity of the disturbance. The height and destructiveness of the waves arriving at any place depend upon its distance from the epicenter, topography of the ocean floor, and the coast line itself. The angle at which the wave arrives, the shape of the coast line, and the topography along the coast and offshore all have their effect. The position of the shore is also a factor, as it may be sheltered by intervening land, or be in a position where waves have a tendency to converge, either because of refraction or reflection, or both.

In addition to seismic sea waves, earthquakes below the surface of the sea may produce a longitudinal wave that travels upward toward the surface, at the speed of sound. When a ship encounters such a wave, it is felt as a sudden shock which may be of such severity that the crew thinks the vessel has struck bottom. Because of such reports, some older charts indicated shoal areas at places where the depth is now known to be a thousand fathoms or more.

Storm Waves—In relatively tideless seas like the Baltic and Mediterranean, winds cause the chief fluctuations in sea level. Elsewhere, the astronomical tide usually masks these variations. However, under exceptional conditions, either severe extratropical storms or tropical cyclones can produce changes in sea level that exceed the normal range of tide. Low sea level is of little concern except to shipping, but a rise above ordinary high-water mark, particularly when it is accompanied by high waves, can result in a catastrophe.

Although, like tsunamis, these **storm waves** or **"storm surges"** are popularly called **tidal waves,** they are not associated with the tide. They consist of a single wave crest and hence have no period or wave length.

Three effects in a storm induce a rise in sea level. The first is wind stress on the sea surface, which results in a piling-up of water (sometimes called "wind set-up"). The second effect is the convergence of wind-driven currents, which elevates the sea surface along the convergence line. In shallow water, bottom friction and the effects of local topography cause this elevation to persist and may even intensify it. The low atmospheric pressure that accompanies severe storms causes the third effect, which is sometimes referred to as the "inverted barometer." An inch of mercury is equivalent to about 13.6 inches of water and the adjustment of the sea surface to the reduced pressure can amount to several feet at equilibrium.

All three of these causes act independently, and if they happen to occur simultaneously, their effects are additive. In addition, the wave can be intensified or amplified by the effects of local topography. Storm waves may reach heights of 20 feet or more, and it is estimated that they cause three-fourths of the deaths attributed to hurricanes.

Standing Waves and Microseisms—Previous sections in this chapter have dealt with **progressive waves** which appear to move regularly with time. When two systems of progressive waves having the same period travel in opposite directions across the same area, a series of **standing waves** may form. These appear to remain stationary. Recent investigation has indicated that when this condition occurs, a pressure variation is exerted on the ocean bottom proportional to the product of the wave heights of the two wave systems. The period of these pressure variations is half that of the progressive waves. The magnitude and period of these variations are of the right order to cause a series of minute earth shocks of the magnitude of those recorded by very sensitive seismographs and known as **microseisms.** It is probable, therefore, that microseisms are generated by standing waves established in any manner, as by waves from independent sources, those in the wake of a moving circulation, waves at the center of a stationary circulation, or by reflection of waves striking a steep shore.

Another type of standing wave, called a **seiche** (sāsh), sometimes occurs in a confined body of water. It is a long wave, usually having its crest at one end of the confined space, and its trough at the other. Its period may be anything from a few minutes to an hour or more, but somewhat less than the tidal period. Seiches are usually attributed to strong winds or differences in atmospheric pressure.

Tide Waves—There are, in general, two regions of high tide separated by two regions of low tide, and these regions move progressively westward around the earth as the moon revolves in its orbit. The high tides are the crests of these **tide waves,** and the low tides are the troughs. The wave is not noticeable at sea, but becomes apparent along the coasts, particularly in funnel-shaped estuaries. In certain river mouths or estuaries of particular configuration, the incoming wave of high water overtakes the preceding low tide, resulting in a high-crested, roaring wave which progresses upstream in one mighty surge called a **bore.**

Internal Waves—Thus far, the discussion has been confined to waves on the surface of the sea, the boundary between air and water. **Internal waves,** or **boundary waves,** are created below the surface, at the boundaries between water strata of different densities. The density differences between adjacent water strata in the sea are considerably less than that between sea and air. Consequently, internal waves are much more easily formed than surface waves, and they are often much larger. The maximum height of wind waves on the surface is about 60 feet, but internal wave heights as great as 300 feet have been encountered.

Internal waves are detected by a number of observations of the vertical temperature distribution, using recording devices such as the bathythermograph. They have periods as short as a few minutes, and as long as 12 or 24 hours, these greater periods being associated with the tides.

A slow-moving ship operating in a fresh water layer having a depth approximating the draft of the vessel may produce short-period internal waves. This may occur off rivers emptying into the sea or in polar regions in the vicinity of melting ice. Under suitable conditions, the normal propulsion energy of the ship is expended in generating and maintaining these internal waves and the ship appears to "stick" in the water, becoming sluggish and making little headway. The phenomenon, known as **dead water,** disappears when speed is increased by a few knots.

The full significance of internal waves has not been determined, but it is known that they may cause submarines to rise and fall like a ship at the surface, and they may also affect sound transmission in the sea.

Waves and Ships—The effects of waves on a ship vary considerably with the type of ship, its course and speed, and the condition of the sea. A short vessel has a tendency to ride up one side of a wave and down the other side, while a larger vessel may tend to ride *through* the waves on an even keel. If the waves are of such length that the bow and stern of a vessel are alternately in successive crests and successive troughs, the vessel is subject to heavy sagging and hogging stresses, and under extreme conditions may break in two. A change of heading may reduce the danger. Because of the danger from sagging and hogging, a small vessel is sometimes better able to ride out a storm than a large one.

If successive waves strike the side of a vessel at the same phase of successive rolls, relatively small waves can cause heavy rolling. The effect is similar to that of swinging a child, where the strength of the push is not as important as its timing. The same effect, if applied to the bow or stern in time with the pitch, can cause heavy pitching. A change of either heading or speed can reduce the effect.

A wave having a length twice that of a ship places that ship in danger of falling off into the trough of the sea, particularly if it is a slow-moving vessel. The effect is especially pronounced if the sea is broad on the bow or broad on the quarter. An increase of speed reduces the hazard.

Use of Oil for Modifying the Effects of Breaking Waves—Oil has proved effective in modifying the effects of breaking waves, and has proved useful to vessels at sea, whether making way or stopped, particularly when lowering or hoisting boats. Its effect is greatest in deep water, where a small quantity suffices if the oil can be made to spread to windward. In shallow water where the water is in motion over the bottom, oil is less effective but of some value.

The heaviest oils, notably animal and vegetable oils, are the most effective. Crude petroleum is useful, but its effectiveness can be improved by mixing it with animal and vegetable oils. Gasoline or kerosene are of little value. Oil spreads slowly. In cold weather it may need some thinning with petroleum to hasten the process and produce the desired spread before the vessel is too far away for the effect to be useful.

At sea, best results can be expected if the vessel drifts or runs slowly before the wind, with the oil being discharged on both sides from waste pipes or by another convenient method. If a sea anchor is used, oil can be distributed from a container inserted within it for this purpose. If such a container is not available, an oil bag can be fastened to an endless line rove through a block on the sea anchor. This permits distribution of oil to windward, and provides a means for hauling the bag aboard for refilling.

If another vessel is being towed, the oil should be distributed from the towing vessel, forward and on both sides, so that both vessels will be benefited. If a drifting vessel is to be approached, the oil might be distributed from both sides of the drifting vessel or by the approaching vessel, which should distribute it to leeward of the drifting vessel so that that vessel will drift into it. If the vessel being approached is aground, the procedure best suiting the circumstances should be used.

If oil is needed in crossing a bar to enter a harbor, it can be floated in ahead of the vessel if a flood current is running. A considerable amount may be needed. During slack water a hose might be trailed over the bow and oil poured freely through it if no more convenient method is available. With ebb current oil is of little use, unless it can be distributed from another vessel or in some other manner from the opposite side of the bar.

=2=
The Oceans

Oceanography is the application of the sciences to the phenomena of the oceans. It includes a study of their forms; physical, chemical, and biological features; and phenomena. It embraces the widely separated fields of geography, geology, chemistry, physics, and biology. Many subdivisions of these sciences, such as sedimentation, ecology (biological relationship between organisms and their environment), bacteriology, biochemistry, hydrodynamics, acoustics, and optics, have been extensively studied in the oceans.

The oceans cover 70.8 percent of the surface of the earth. The Atlantic covers 16.2 percent, the Pacific 32.4 percent (3.2 percent more than the land area of the entire earth), the Indian Ocean 14.4 percent, and marginal and adjacent areas (of which the largest is the Arctic Ocean) 7.8 percent. Their extent alone makes them an important subject for study. However, greater incentive lies in their use for transportation, their influence upon weather and climate, and their potentiality as a source of power, food, fresh water, and mineral and organic substances.

History of Oceanography—The earliest studies of the oceans were concerned principally with problems of navigation. Information concerning tides, currents, soundings, ice, and distances between ports was needed as ocean commerce increased. According to Posidonius, a depth of 1,000 fathoms had been measured in the Sea of Sardinia as early as the second century BC. About the middle of the 19th century, the Darwinian theories of evolution gave a great impetus to the collection of marine organisms, since it is believed by some that all terrestrial forms have evolved from oceanic ancestors. Later, the serious depletion of many fisheries called for investigation of the relation of the economically valuable organisms to the physical characteristics of their environment, especially in northwestern Europe and off Japan. Still later, the growing use of the oceans in warfare, particularly after the development of the submarine, required that much effort be expended in problems of detection and attack, resulting in the study of many previously neglected scientific aspects of the sea.

Exploration of the seas was primarily geographical until the 19th century, although the accumulated observations of seafarers, as recorded in the early charts and sailing directions, often included data on tides, currents, and other oceanographic phenomena. The great voyages of discovery, particularly those beginning in 1768 with Captain Cook, and continued by such commanders as La Perouse, Bellingshausen, and Wilkes, included scientists in their complements. However, scientific work on the oceans at this period was severely limited by lack of suitable instruments for probing conditions below the surface. Meanwhile, Lieutenant Matthew Fontaine Maury, USN, working in the forerunner of the U. S. Navy Hydrographic Office in Washington, developed to a high degree of perfection the analysis of log-book observations. His first results, published in 1848, were of great importance to ship operations in the recommendation of favorable sailing routes, and they stimulated international cooperation in the fields of oceanography and marine meteorology.

In the rapid advances in technology after 1850, oceanographic instrumentation problems were not neglected, with the result that the British Navy in 1872–76 was able to send HMS *Challenger* around the world on the first purely deep-sea oceanographic expedition ever attempted. Her bottom samples, as analyzed by Sir John Murray, laid the foundation of geological oceanography, and 77 of her sea water samples, analyzed by C. R. Dittmar, proved for the first time that various constituents of the salts in sea water are everywhere in virtually the same proportions.

Since that time, the coastal waters and fishing banks of many nations have been extensively studied, and numerous vessels of various nationalities have conducted work on the high seas. Notable among these have been the American *Albatross* from 1882 to 1920; the Austrian *Pola* in the Mediterranean and Red Seas between 1890 and 1896; the Danish *Dana*, which during its voyages of 1920–22 discovered the breeding place of the European eels in the Sargasso Sea; the American *Carnegie* in 1927–29; the German *Meteor* in the Atlantic from 1928 to 1938; and the British *Discovery II* in the antarctic between 1930 and 1939. Notable also were the drifts of the Norwegian vessels *Fram* and *Maud* in the artic ice pack from 1893 to 1896 and 1918 to 1925, respectively; the attempt by Sir George Hubert Wilkins to operate under the ice in the British submarine *Nautilus* in 1931; and the Russian station set up at the north pole in 1937, which made observations from the drifting pack ice.

At the same time, investigations pursued ashore provided the theoretical basis for the explanation of ocean currents, under the leadership of Helland-Hansen in Norway and Ekman and the Bjerknes in Sweden, while Martin Knudsen in Denmark worked out the precise details of the relationship between chlorinity, salinity, and density, enabling the theories to be verified by field observations.

During World War II, basic investigations were interrupted while work on purely military applications of oceanography was carried out. Deep-sea expeditions were renewed by the Swedish *Albatross* after the war, followed by the Danish *Galathea*, the second British *Challenger* (built in 1931) and *Discovery II* in the antarctic, and vessels of the American Scripps Institution in the Pacific. Oceanographic work was carried out by Americans and Russians in the arctic. By 1961, a total of ten Russian and three United States drifting ice stations had been established. Two United States stations were also established aboard floating ice islands.

Among the leading oceanographic institutions in Europe are the Geophysical Institute of the University of Bergen in Norway; the Oceanographic Institute at Göteborg, Sweden; the National Institute of Oceanography in Great Britain; the German Hydrographic Institute in Hamburg; and the Museum of Oceanography at Monaco. The Marine Biological Station at Naples, Italy, has served as a model for others throughout the world.

In the Far East, the Hydrographic Division of the Maritime Safety Agency is perhaps the most prominent of a number of Japanese oceanographic activities. The Institute of Oceanology at Vladivostok is the foremost oceanographic establishment on the Asiatic mainland.

Canada maintains the Pacific Oceanographic Group at Nanaimo, B. C., and the Atlantic Oceanographic Group at St. Andrews, N. B. In the United States, the leading nongovernmental oceanographic institutions include the Scripps Institution of Oceanography of the University of California, La Jolla, Calif.; the Department of Oceanography of the University of Washington, Seattle, Wash.; Woods Hole Oceanographic Institution, Woods Hole, Mass.; the Marine Laboratory of the University of Miami, Coral Gables, Fla.; and the Department of Oceanography of Texas A. & M. College, College Station, Tex.

There exist also various international organizations in the field of oceanography, which coordinate and promote international cooperation. The International Council for the Exploration of the Sea, with headquarters in Copenhagen, which was established to exchange data on fisheries problems in the waters of northwestern Europe, has been notably successful, and similar organizations have been established in other areas.

Origin of the Oceans—Although many leading geologists still disagree with the conclusion that the structure of the continents is fundamentally different from that of the oceans, there is a growing body of evidence in support of the theory that the rocks underlying the ocean floors are more dense than those underlying the continents. According to this theory, all the earth's crust floats on a central liquid core, and the portions that make up the continents, being lighter, float with a higher freeboard. Thus, the thinner areas, composed of heavier rock, form natural basins where water has collected.

The origin of the water in the oceans is also controversial. Although some geologists have postulated that all the water existed as vapor in the atmosphere of the primeval earth, and that it fell in great torrents of rain as soon as the earth cooled sufficiently, another school holds that the atmosphere of the original hot earth was lost, and that the water gradually accumulated as it was given off in steam by volcanoes or worked to the surface in hot springs.

Most of the water on the earth's crust is now in the oceans—about 328,000,000 cubic statute miles, or about 85 percent of the total. The mean depth of the ocean is 2,075 fathoms, and the total area is 139,000,000 square statute miles.

Oceanographic Chemistry may be divided into three main parts: the chemistry of (1) sea water, (2) marine sediments, and (3) organisms living in the sea. The first is of particular interest to the navigator.

Chemical properties of sea water are determined by analyzing samples of water obtained at various places and depths. Samples from below the surface are obtained by means of metal bottles designed for this purpose. The open bottles are attached at suitable intervals to a wire lowered into the sea. When they reach the desired depths, a metal ring or **messenger** is dropped down the wire. When the messenger arrives at the first bottle, it causes the bottle to close, trapping a sample of the water at that depth, and releasing a second messenger which travels on down the wire. The process is repeated at each bottle until all are closed, when they are hauled up and each bottle detached as it comes within reach. Of the various types devised, the **Nansen bottle** is the most widely used. It is equipped with a removable frame for attaching a thermometer.

For centuries table salt has been produced from sea water by natural evaporation in countries with a suitable climate. More recently, practical industrial processess have been developed for recovering bromine and magnesium from the sea. Calcium carbonate, in the form of oyster shells or coral rock, is obtained after precipitation by living organisms.

Three elements in the sea, silicon, nitrogen, and phosphorus, are most significant in the growth of living organisms.

Certain of the elements, notably chlorine, bromine, sulfur, and boron, are much more abundant in the ocean than in the rest of the earth's crust. These elements are among the more volatile ones, and their abundance in the sea tends to confirm the hypothesis that volcanic action is largely responsible for the present oceans.

In many cases, chemical relationships influence the abundance of elements in the sea. Barium, for example, forms a sulfate of very limited solubility, and thus the high concentration of sulfate in sea water limits the possible amount of dissolved barium. Thus, the concentration of many elements is limited by the solubility of their most insoluble compounds.

In addition to dissolved solids, sea water contains in solution all of the gases found in the atmosphere, but not in the same proportions. The most abundant is nitrogen, which, however, because of its chemical inertness, does not enter into biological processes. Oxygen, produced in the surface layers by plant photosynthesis or dissolved directly from the atmosphere, is of major importance for all forms of life. By biological activity, the oxygen concentration at depths below the surface is usually reduced to a fraction of the surface values, and under certain conditions, owing either to the presence of abundant oxidizable material, or a stagnant condition, or both, it may become completely exhausted. Under these conditions, sulfate-reducing bacteria produce hydrogen sulfide gas from the abundant sulfate in sea water. The existence of such conditions is often indicated to the mariner by the blackening of white lead paint, a well-known phenomenon in badly polluted estuaries.

Hydrogen sulfide may also be encountered at great depths in the ocean. The fiords of Norway, deep channels cut by former glaciers, are characterized in general by shallow sills at the entrances, where the terminal moraines of the glaciers were deposited. These sills serve as barriers to the mixing and renewing of the deeper waters within the fiords, and, as a result, conditions producing hydrogen sulfide are frequently encountered.

A similar situation exists in the Black Sea. Here the Bosporus and Dardanelles act as sills, and all the deeper water of the Black Sea is cut off from contact with the surface waters, which, diluted by the runoff from the Danube and Don Rivers, have a salinity of about 17.5 parts per thousand. The deeper water, renewed only by the bottom current through the Bosporus, has a salinity of 22 parts per thousand, and the great density difference between the surface layers and the deeper water effectively prevents mixing and the transfer of dissolved oxygen from the surface layers to greater depths. Below about 100 fathoms, therefore, the waters of the Black Sea are completely devoid of dissolved oxygen, containing instead large concentrations of hydrogen sulfide.

No living creatures exist under these conditions except anaerobic bacteria, which comprise the only form of life in five-sixths of the waters of the Black Sea.

Physical properties of sea water are dependent primarily upon salinity, temperature, and pressure. Factors such as motion of the water and the amount of suspended matter affect such properties as color and transparency, conduction of heat, absorption of radiation, etc.

Salinity is the amount of dissolved solid material in the water, usually expressed as parts per thousand (by weight), under certain standard conditions. This is not the same as **chlorinity,** which is equal approximately to the amount of chlorine in the water. (Actually the chlorine content is about 1.00045 times the chlorinity as determined by standard procedures.) The two have been found to be related empirically by the formula

$$\text{salinity} = 0.03 + 1.805 \times \text{chlorinity}.$$

Since the determination of salinity is a slow and difficult process, while chlorinity can be determined easily and accurately by titration with silver nitrate, it is customary to determine chlorinity and compute salinity by the formula given above. By this process, salinity can be determined with an error not exceeding 0.02 parts per thousand. It generally varies between about 33 and 37 parts per thousand, the average being about 35 parts per thousand. However, when the water has been diluted, as near the mouth of a river or after a heavy rainfall, the salinity is somewhat less; and in areas of excessive evaporation, the salinity may be as high as 40 parts per thousand. In certain confined bodies of water, notably the Great Salt Lake in Utah, and the Dead Sea in Asia Minor, the salinity is several times this maximum. Chlorinity accounts for about 55 percent of salinity, the average being about 19 parts per thousand.

Temperature in the ocean varies widely, both horizontally and with depth. Maximum values of about 90° F are encountered in the Persian Gulf in summer, and the lowest possible values of about 28° F (the usual minimum freezing point of sea water) occur in polar regions.

The vertical distribution of temperature in the sea nearly everywhere shows a decrease of temperature with depth. Since colder water is denser, it sinks below warmer water. This results in a temperature distribution just opposite to that in the earth's crust, where temperature increases with depth below the surface of the ground.

In general, in the sea there is usually a mixed layer of isothermal water below the surface, where the temperature is the same as that of the surface. This layer is best developed in the trade-wind belts, where it may extend to a depth of 100 fathoms; in temperate latitudes in the spring, it may disappear entirely. Below this layer is a zone of rapid temperature decrease, called the **thermocline,** to the temperature of the deep oceans. At a depth greater than 200 fathoms, the temperature everywhere is below 60° F, and in the deeper layers, fed by cooled waters that have sunk from the surface in the arctic and antarctic, temperatures as low as 33° F exist.

In the deepest ocean basins, the temperature increases slightly with depth, the increase being about 1° F at 3,000 fathoms. The warming is believed to be caused more by the slight compression of sea water than by heat from the earth's crust.

Pressure—In oceanographic work, pressure is generally expressed in units of the centimeter-gram-second system. The basic unit of this system is one dyne per square centimeter. This is a very small unit, one million constituting a practical unit called a bar, which is nearly equal to one atmosphere. Atmospheric pressure is often expressed in terms of **millibars,** 1,000 of these being equal to one bar. In oceanographic work, water pressure is commonly expressed in terms of **decibars,** ten of these being equal to one bar. One decibar is equal to nearly 1½ pounds per square inch. This unit is convenient because it is very nearly the pressure exerted by one meter of water. Thus, the pressure in decibars is approximately the same as the depth in meters, the unit of depth customarily used in oceanographic research. In terms more familiar to the mariner, the pressure at various depths is as follows:

Depth in fathoms	Pressure in Pounds per square inch
1,000	2,680
2,000	5,390
3,000	8,100
4,000	10,810
5,000	13,520

The increase in pressure with depth is nearly constant because water is only slightly compressible.

Although virtually all of the physical properties of sea water are affected to a measurable extent by pressure, the effect is not as great as those of salinity and temperature. Pressure is of particular importance to submarines, directly because of the stress it induces in the materials of the craft, and indirectly because of its effect upon buoyancy.

Density is mass per unit volume. Oceanographers use the centimetergram-second system, in which density is expressed as grams per cubic centimeter. The ratio of the density of a substance to that of a standard substance under stated conditions is called **specific gravity.** By definition, the density of distilled water at $4°$ C ($39°.2$F) is one gram per milliliter (approximately one gram per cubic centimeter). Therefore, if this is used as the standard, as it is in oceanographic work, density and specific gravity are virtually identical numerically.

The density of sea water depends upon salinity, temperature, and pressure. At constant temperature and pressure, density varies with salinity or, because of the relationship between this and chlorinity, with the chlorinity. A temperature of $32°$ F and atmospheric pressure are considered standard for density determination. The effects of thermal expansion and compressibility are used to determine the density at other temperatures and pressures. The density at a particular pressure affects the buoyancy of submarines. It is also important in its relation to ocean currents.

The greatest changes in density of sea water occur at the surface, where the water is subject to influences not present at depths. Here density is decreased by precipitation, run-off from land, melting of ice, or heating. When the surface water becomes less dense, it tends to float on top of the more dense water below. There is little tendency for the water to mix, and so the condition is one of stability. The density of surface water is increased by evaporation, formation of sea ice, and by cooling. If the surface water becomes more dense than that below, it sinks to the level at which other water has the same density. Here it tends to spread out to form a layer, or to increase the thickness of the layer below it. The less dense water rises to make room for it, and the surface water moves in to replace that which has descended. Thus, a convective circulation is established. It continues until the density becomes uniform from the surface to the depth at which a greater density occurs. If the surface water becomes sufficiently dense, it sinks all the way to the bottom. If this occurs in an area where horizontal flow is unobstructed, the water which has descended spreads to other regions, creating a dense bottom layer. Since the greatest increase in density occurs in polar regions, where the air is cold and great quantities of ice form, the cold, dense polar water sinks to the bottom and then spreads to lower latitudes. This process has continued for a sufficiently long period of time that the entire ocean floor is covered with this dense polar water, thus explaining the layer of cold water at great depths in all the oceans.

Compressibility—Sea water is nearly incompressible, its coefficient of compressibility being only 0.000046 per bar under standard conditions. This value changes slightly with changes of temperature or salinity. The effect of compression is to force the molecules of the substance closer together, causing it to become more dense. Even though the compressibility is low, its total effect is considerable because of the amount of water involved. If the compressibility of sea water were zero, sea level would be about 90 feet higher than it now is.

Viscosity is resistance to flow. Sea water is slightly more viscous than fresh water. Its viscosity increases with greater salinity, but the effect is not nearly as marked as that occurring with decreasing temperature. The rate is not uniform, becoming greater as the temperature decreases. Because of the effect of temperature upon viscosity, an incompressible object might sink at a faster rate in warm surface water than in colder water below. However, for most objects, this effect may be more than offset by the compressibility of the object.

The actual relationships existing in the ocean are considerably more complicated than indicated by the simple explanation given above, because of turbulent motion within the sea. The disturbing effect is called **eddy viscosity**.

Transparency of sea water varies with the number, size, and nature of particles suspended in the water, as well as with the nature and intensity of illumination. The rate of decrease of light energy with depth is called the "extinction coefficient." The earliest method of measuring transparency was by means of a **Secchi disk,** a white disk 30 centimeters (a little less than one foot) in diameter. This was lowered into the sea, and the depth at which it disappeared was recorded. In coastal waters the depth varies from about 5 to 25 meters (16 to 82 feet). Offshore, the depth is usually about 45 to 60 meters (148 to 197 feet). The greatest recorded depth at which the disk has disappeared is 66 meters (217 feet), in the Sargasso Sea.

Although the Secchi disk still affords a simple method of measuring transparency, more exact methods have been devised.

Color—The color of sea water varies considerably. Water of the Gulf Stream is a deep indigo blue, while a similar current off Japan was named Kuroshio (Black Stream) because of the dark color of its water. Along many coasts the water is green. In certain localities a brown or brownish-red water has been observed.

Offshore, some shade of blue is common, particularly in tropical or sub-tropical regions. It is due to scattering of sunlight by minute particles suspended in the water, or by molecules of the water itself. Because of its short wave length, blue light is more effectively scattered than light of longer waves. Thus, the ocean appears blue for the same reason that the sky does. The green color often seen near the coast is a mixture of the blue due to scattering of light and a stable soluble yellow pigment associated with phytoplankton. Brown or brownish-red water receives its color from large quantities of certain types of **algae,** microscopic plants in the sea.

Marine geology is a branch of oceanography dealing with bottom relief, particularly the characteristics of ocean basins and the geological processes that brought them into being and tend to alter them, as well as with marine sediments.

Bottom Relief—Compared to land, relatively little is known of relief below the surface of the sea. It would be difficult to withhold knowledge of a major land feature in an area often visited by man, but the sea has until recent years proved an effective barrier to acquisition of knowledge of features below its surface. Although soundings of 1,000 fathoms were probably made as early as the second century BC (art. 3002), the number of deep sea soundings by means of a weight lowered to the bottom has been relatively few. The process is a time-consuming one requiring special equipment. Several hours are needed for a single sounding. Since the development of an effective echo sounder in 1922, the number of deep sea soundings has greatly increased. Later, a recording echo sounder was developed to permit the continuous tracing of a **bottom profile.** This has assisted materially in the acquisition of knowledge of bottom relief. By this means, many flat-topped seamounts (called **guyots**), mountain ranges, and other features have been discovered. Although the main features are becoming known, a great many details are yet to be learned.

Along most of the coasts of the continents, the bottom slopes gradually downward to a depth of about 100 fathoms or somewhat less, where it falls away more rapidly to greater depths. The **continental shelf** averages about 30 miles in width, but varies from nothing to about 800 miles, the widest part being off the Siberian arctic coast. A similar shelf extending outward from an island or group of islands is called an **insular shelf.** At the outer edge of the shelf, the steeper slope of 2° to 4° is called the **continental talus** or **continental slope,** or the **insular talus** or **insular slope,** according to whether it surrounds a continent or group of islands. The shelf itself is not uniform, but has numerous hills, ridges, terraces, and canyons, the largest being comparable in size to the Grand Canyon.

As a general rule, the slope of the deep sea bottom is gradual, averaging between 20′ and 40′, but there are many exceptions to this. Off a volcanic island it may be as much as 45°. The relief of the ocean floor is comparable to that of land. Both have steep, rugged mountains, deep canyons, rolling hills, plains, etc. Most of the ocean floor is considered to be made up of a number of more-or-less circular or oval depressions called **basins,** surrounded by walls of lesser depth.

The average depth of water in the oceans is 2,075 fathoms (12,450 feet), as compared to an average height of land above the sea of about 2,750 feet. The greatest known depth is 35,640 feet, in the Marianas Trench in the Pacific. The highest known land is Mount Everest, 29,002 feet. About 23 percent of the ocean is shallower than 10,000 feet, about 76 percent is between 10,000 and 20,000 feet, and a little more than one percent is deeper than 20,000 feet. A very deep part, generally that below 3,000 fathoms, is called a **deep.** A long, narrow depression with steep sides is called a **trench.**

Marine Sediments—The ocean floor is composed of material deposited there through the years. This material consists principally of (1) earth and rocks washed into the sea by streams and waves, (2) volcanic ashes and lava, and (3) the remains of marine organisms. Lesser amounts of land material are carried into the sea by glaciers, or blown out to sea by wind. In the ocean, the material is transported by ocean currents, waves, and ice. Near shore the material is deposited at the rate of about three inches in 1,000 years, while in the deep water offshore the rate is only about half an inch in 1,000 years. Marine deposits in water deep enough to be relatively free from wave action are subject to little erosion. Because of this and the slow rate of deposit, marine sediments provide a better geological record than does the land.

Marine sediments are composed of individual particles of all sizes from the finest clay to large boulders. In general, the inorganic deposits near shore are relatively coarse (sand, gravel, shingle, etc.), while those in deep water are much finer (clay). In some areas the siliceous remains of marine organisms or the calcareous deposits (of either organic or inorganic origin) are sufficient to predominate on the ocean floor.

A wide range of colors is found in marine sediments. The lighter colors (white or a pale tint) are usually associated with coarse-grained quartz or limestone deposits. Darker colors (red, blue, green, etc.) are usually found in mud having a predominance of some mineral substance, such as an oxide of iron or manganese. Black mud is often found in an area that is little disturbed, such as at the bottom of an inlet or in a depression without free access to other areas.

Marine sediments are studied primarily by means of bottom samples. Samples of surface deposits are obtained by means of a **snapper** (for mud, sand, etc.) or "dredge" (usually for rocky material). If a sample of material below the bottom surface is desired, a "coring" device is used. This device consists essentially of a tube driven into the bottom by weights or explosives. A sample obtained in this way preserves the natural order of the various layers. Samples of more than 100 feet in depth have been obtained by means of coring devices. The bottom sample obtained by the mariner, by arming his lead with tallow or soap, is an incomplete indication of bottom surface conditions.

Marine Biology—Sea water has all of the chemical elements needed to sustain plant and animal life. Because of this, and the fact that the oceans contain about 300 times as much space for the existence of life as is available on land and in fresh water, organic material is present in vast quantities.

Marine life may be divided into three major groups: (1) **nekton** (strong-swimming animals such as fish), (2) **plankton** (tiny floating plants or feebly swimming or floating animals), and (3) **benthos** (plants and animals living on the bottom, such as seaweed, barnacles, and crabs). Plankton may be divided into: (a) the **phytoplankton,** consisting of microscopic floating plants; and (b) the **zooplankton,** consisting of feebly swimming or floating animals. Most plankton vary in size from microscopic units to those a small fraction of an inch in length.

Most organic material in the sea is in the form of plankton, which is carried by the ocean currents, not having sufficient strength to choose its environment. Either directly or indirectly, nearly all marine life depends upon these organisms. By means of **photosynthesis,** a process using sunlight, phytoplankton changes chemical nutrients (silicates, nitrates, phosphates) in the sea into primary food which is used by the zooplankton and, to some extent, by larger animals. However, most of the larger animals feed upon the zooplankton. The chemical nutrients are replaced by the excretion of animals and bacterial action in the decomposition of dead plants and animals. Thus, a food cycle is continually going on from chemical nutrient to phytoplankton to zooplankton to nekton and benthos to chemical nutrients.

As indicated above, growth of phytoplankton requires both sunlight and a supply of chemical nutrients. Sunlight in sufficient strength to permit photosynthesis penetrates to a maximum depth of about 500 feet or less. This upper layer in which the process occurs is called the **euphotic zone.** Within this zone, photosynthesis is limited primarily by the supply of chemical nutrients. Under favorable conditions, phytoplankton may increase by as much as 300 percent in a single day.

The abundance of marine life is directly related to the supply of phytoplankton. In shallow water, the chemical nutrients on the bottom are stirred up by motion of the water, and carried into the euphotic zone. This is why an area such as the Grand Banks is a good fishing ground. In polar regions the chemical nutrients are relatively abundant, being brought to the surface by convective currents as the cold surface water sinks and is replaced by the warmer water from the bottom. In the tropics, on the other hand, the sea is relatively stable, and the chemical nutrients have a tendency to sink below the euphotic zone. Even though the clear, blue water has the deepest euphotic zone, photosynthesis proceeds at a slow rate. For this reason blue is sometimes called the "desert color of the sea."

Ocean currents and marine life are so interrelated that currents can sometimes be traced by their supply of plankton. In general, the oceanic circulation helps sustain marine life by stirring up the chemical nutrients and carrying them, or the plankton formed from them, into regions which have an inadequate supply. However, the reverse effect can occur. A notable example occurs from time to time off the west coast of South America. At varying intervals averaging about 12 years, a well-developed stream of tropical water having a relatively small supply of chemical nutrients and plankton flows southward, close to the shore. This water replaces the colder water which is rich in chemical nutrients and plankton. The result is a wholesale destruction of fish which cannot obtain a sufficient supply of food. In some areas the dead fish are washed ashore in such quantities as to constitute a serious problem. With the destruction of so many fish, the supply of guano also decreases because of the death of large numbers of the birds which depend upon the fish for their food supply. Since it commonly occurs near Christmas, this phenomenon is called "El Niño." A strong current such as the Gulf Stream annually carries many fish to their deaths by transporting them from their normally warm habitat to areas where they encounter water which is too cold for them to endure.

=3=
Surf

Surf and the navigator—The prospect of grounding in breaking surf is a nightmare to a deepwater navigator. The purpose of this chapter is to acquaint the navigator with the oceanographic factors affecting breakers and surf.

Refraction—As explained in Chapter 1, wave speed is slowed in shallow water, causing **refraction** if the waves approach the beach at an angle. Along a perfectly straight beach, with uniform shoaling, the wave fronts tend to become parallel to the shore. Any irregularities in the coast line or bottom contours affect the refraction, causing irregularity. In the case of a ridge perpendicular to the beach, for instance, the shoaling is more rapid, causing greater refraction. The waves tend to align themselves with the bottom contours. Waves on both sides of the ridge have a component of motion toward the ridge. This **convergence** of water toward the ridge causes an increase in wave or breaker height. A submarine canyon or valley perpendicular to the beach, on the other hand, produces **divergence,** with a decrease in wave or breaker height. These effects are illustrated in figure 3–1. Bends in the coast line have a similar effect, convergence occurring at a *point*, and divergence if the coast is concave to the sea.

Under suitable conditions, currents also cause refraction. This is of particular importance at entrances of tidal estuaries. When waves encounter a current running in the opposite direction, they become higher and shorter. This results in a choppy sea, often with breakers. When waves move in the same direction as current, they decrease in height, and become longer. Refraction occurs when waves encounter a current at an angle.

Breakers and Surf—In deep water, swell generally moves across the surface as somewhat regular, smooth undulations. When shoal water is reached, the wave period remains the same, but the speed

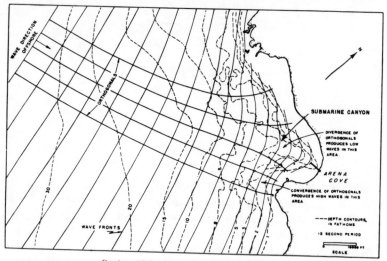

Courtesy of Robert L. Wiegel, Council on Wave Research, University of California.

Figure 3–1 *The effect of bottom topography in causing wave convergence and wave divergence.*

decreases. The amount of decrease is negligible until the depth of water becomes about one-half the wave length, when the waves begin to "feel" bottom. There is a slight decrease in wave height, followed by a rapid increase, if the waves are traveling perpendicular to a straight coast with a uniformly sloping bottom. As the waves become higher and shorter, they also become steeper, and the crest becomes narrower. When the speed of individual particles at the crest becomes greater than that of the wave, the front face of the wave becomes steeper than the rear face. This process continues at an accelerating rate as the depth of water decreases. At some point the wave may become unstable, toppling forward to form a **breaker.**

There are three general classes of breakers. A **spilling breaker** breaks gradually over a considerable distance. A **plunging breaker** extends to curl over and break with a single crash. A **surging breaker** peaks up, but surges up the beach without spilling or plunging. It is classed as a breaker even though it does not actually break. The type of breaker is determined by the steepness of the beach and the steepness of the wave before it reaches shallow water, as illustrated in figure 3–2.

Longer waves break in deeper water, and have a greater breaker

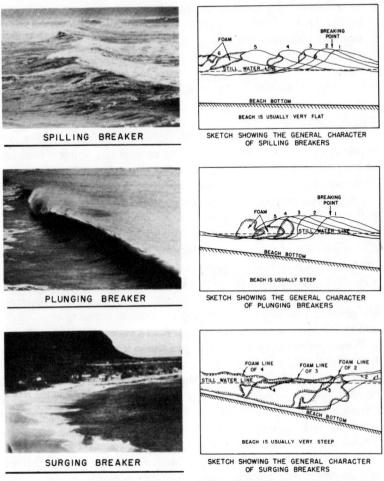

SPILLING BREAKER

SKETCH SHOWING THE GENERAL CHARACTER OF SPILLING BREAKERS

PLUNGING BREAKER

SKETCH SHOWING THE GENERAL CHARACTER OF PLUNGING BREAKERS

SURGING BREAKER

SKETCH SHOWING THE GENERAL CHARACTER OF SURGING BREAKERS

Courtesy of Robert L. Weigel, Council on Wave Research, University of California.

Figure 3–2 *The three types of breakers.*

height. The effect of a steeper beach is also to increase breaker height. The height of breakers is less if the waves approach the beach at an acute angle. With a steeper beach slope there is greater tendency of the breakers to plunge or surge. Following the **uprush** of water onto a beach after the breaking of a wave, the seaward **backrush** occurs. This tends to further slow the bottom of a wave, thus increasing its tendency to break. This effect is greater as either the speed or depth of the backwash increases. The still water depth at the point of breaking is approximately 1.3 times the average breaker height.

Surf varies with both position along the beach and time. A change in position often means a change in bottom contour, with the refraction effects discussed above. At the same point, the height and period of waves vary considerably from wave to wave. A group of high waves is usually followed by several lower ones. Therefore, passage through an inlet can usually be made most easily immediately following a series of higher waves.

Since surf conditions are directly related to height of the waves approaching a beach, and the configuration of the bottom, the state of the surf at any time can be predicted if one has the necessary information and knowledge of the principles involved. Height of the sea and swell can be predicted from wind data, and information on bottom configuration can generally be obtained from the nautical chart. In addition, the area of lightest surf along a beach can be predicted if details of the bottom configuration are available. Detailed information on prediction of surf conditions is given in H.O. Pub. No. 234, *Breakers and Surf; Principles in Forecasting.*

Currents in the Surf Zone—In and adjacent to a surf zone, currents are generated by waves approaching the bottom contours at an angle, and by irregularities in the bottom.

Waves approaching at an angle produce a **longshore current** parallel to the beach, within the surf zone. Longshore currents are most common along straight beaches. Their speeds increase with increasing breaker height, decreasing wave period, increasing angle of breaker line with the beach, and increasing beach slope. Speed seldom exceeds one knot, but sustained speeds as high as three knots have been recorded. Longshore currents are usually constant in direction.

As explained earlier, wave fronts advancing over nonparallel bottom contours are refracted to cause convergence or divergence of the energy of the waves. Energy concentrations, in areas of convergence, form barriers to the returning backwash, which is deflected *along* the beach to areas of less resistance. Backwash accumulates at weak points, and returns seaward in concentrations, forming **rip currents** through the surf. At these points the large volume of returning water has a retarding effect upon the incoming waves, thus adding to the condition causing the rip current. The waves on one or both sides of the rip, having greater energy and not being retarded by the concentration of backwash, advance faster and farther up the beach. From here, they move *along* the beach as **feeder currents.** At some

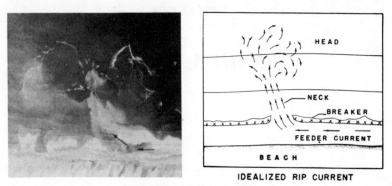

Courtesy of Robert L. Weigel, Council on Wave Research, University of California.

Figure 3–3 *A rip current (left) and a diagram of its parts (right).*

point of low resistance, the water flows seaward through the surf, forming the **neck** of the rip current. Outside the breaker line the current widens and slackens, forming the **head.** The various parts of a rip current are shown in figure 3–3.

Rip currents may also be caused by irregularities in the beach face. If a beach indentation causes an uprush to advance farther than the average, the backrush is delayed and this in turn retards the next incoming **foam line** (the front of a wave as it advances shoreward after breaking) at that point. The foam line on each side of the retarded point continues in its advance, however, and tends to fill in the retarded area, producing a rip current.

=4=

Ocean Currents

The movement of water comprising the oceans is one of the principal sources of discrepancy between dead reckoning and actual positions of vessels. Water in essentially horizontal motion is called a **current,** the direction *toward* which it moves being the **set,** and its speed the **drift.** A well-defined current extending over a considerable region of the ocean is called an **ocean current.**

A **periodic current** is one the speed or direction of which changes cyclically at somewhat regular intervals, as a tidal current. A **seasonal current** is one which has large changes in speed or direction due to seasonal winds. A **permanent current** is one which experiences relatively little periodic or seasonal change.

A **coastal current** flows roughly parallel to a coast, outside the surf zone, while a **longshore current** is one parallel to a shore, inside the surf zone, and generated by waves striking the beach at an angle. Any current some distance from the shore may be called an **offshore current,** and one close to the shore an **inshore current.**

A **surface current** is one present at the surface, particularly one that does not extend more than a relatively few feet below the surface. A **subsurface current** is one which is present below the surface only.

There is evidence to indicate that the strongest ocean currents consist of relatively narrow, high-speed streams that follow winding, shifting courses. Often associated with these currents are secondary **countercurrents** flowing adjacent to them but in the opposite direction, and somewhat local, roughly circular, **eddy currents.** A relatively narrow, deep, fast-moving current is sometimes called a **stream current,** and a broad, shallow, slow-moving one a **drift current.**

Causes of Ocean Currents—Although our knowledge of the processes which produce and maintain ocean currents is far from complete, we do have a general understanding of the principal factors involved. The primary generating force is wind, and the chief secondary force is the density differences in the water. In addition, such factors as depth of water, underwater topography, shape of the basin in which the current is running, extent and location of land, and deflection by the rotation of the earth all affect the oceanic circulation.

Wind Currents—The stress of wind blowing across the sea causes the surface layer of water to move. This motion is transmitted to each succeeding layer below the surface, but due to internal friction within the water, the rate of motion decreases with depth. The current thus set up is called a **wind current.** Although there are many variables, it is generally true that a steady wind for about 12 hours is needed to establish such a current.

A wind current does not flow in the direction of the wind, being deflected by Coriolis force, due to rotation of the earth. This deflection is toward the *right* in the northern hemisphere, and toward the *left* in the southern hemisphere. The Coriolis force is greater in higher latitudes, and is more effective in deep water. In general, the difference between wind direction and surface wind-current direction varies from about 15° along shallow coastal areas to a maximum of 45° in the deep oceans. The angle increases with depth. At several hundred fathoms the current may flow in the opposite direction to the surface current.

The speed of the current depends upon the speed of the wind, its constancy, the length of time it has blown, and other factors. In general, about two percent of the wind speed, or a little less, is a good average for deep water where the wind has been blowing steadily for at least 12 hours.

Currents Related to Density Differences—The density of water varies with salinity, temperature, and pressure. At any given depth, the differences in density are due to differences in temperature and salinity. When suitable information is available, a map showing geographical density distribution at a certain depth could be drawn, with lines connecting points of equal density. These **isopycnic lines,** or lines connecting points at which a given density occurs at the same depth, would be similar to isobars on a weather map, and would serve an analogous purpose, showing areas of high density and those of low density. In an area of high density, the water surface is lower than in an area of low density, the maximum difference in height being of the order of one to two feet in 40 miles. Because of this difference, water tends to flow from an area of higher water (low density) to one of lower water (high density), but due to rotation of the earth, it is deflected toward the right in the northern hemisphere, and toward the left in the southern hemisphere. Thus, a circulation is set up similar to the cyclonic and anticyclonic circulation in the atmosphere. The greater the density gradient (rate of change with distance), the faster the related current.

Oceanic Circulation—A number of ocean currents flow with great persistence, setting up a circulation that continues with relatively little change throughout the year. Because of the influence of wind in creating current, there is a relationship between this oceanic circulation and the general circulation of the atmosphere. The oceanic circulation is shown in figure 4–1, with the names of the major ocean currents. Some differences in opinion exist regarding the names and limits of some of the currents, but those shown are representative. The spacing of the lines is a general indication of speed, but conditions vary somewhat with the season. This is particularly noticeable in the Indian Ocean and along the South China coast, where currents are influenced to a marked degree by the monsoons.

Atlantic Ocean Currents—The trade winds, which blow with great persistence, set up a system of **equatorial currents** which at times extends over as much as 50° of latitude, or even more. There are two westerly flowing currents conforming generally with the areas of trade winds, separated by a weaker, easterly flowing countercurrent.

The **north equatorial current** originates to the northward of the Cape Verde Islands and flows almost due west at an average speed of about 0.7 knot.

The **south equatorial current** is more extensive. It starts off the west coast of Africa, south of the Gulf of Guinea, and flows in a generally westerly direction at an average speed of about 0.6 knot. However, the speed gradually increases until it may reach a value of 2.5 knots or more off the east coast of South America. As the current approaches Cabo de São Roque, the eastern extremity of South America, it divides, the southern part curving toward the south along the coast of Brazil, and the northern part being deflected by the continent of South America toward the north.

Between the north and south equatorial currents a weaker **equatorial countercurrent** sets toward the east in the general vicinity of the doldrums. This is fed by water from the two westerly flowing equatorial currents, particularly the south equatorial current. The extent and strength of the equatorial countercurrent changes with the seasonal variations of the wind. It reaches a maximum during July

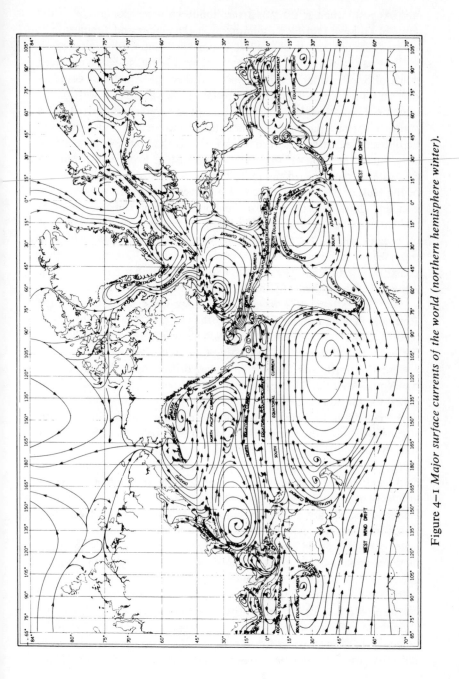

Figure 4–1 *Major surface currents of the world (northern hemisphere winter).*

and August, when it extends from about 50° west longitude to the Gulf of Guinea. During its minimum, in December and January, it is of very limited extent, the western portion disappearing altogether.

That part of the south equatorial current flowing along the northern coast of South America which does not feed the equatorial counter-current unites with the north equatorial current at a point west of the equatorial countercurrent. A large part of the combined current flows through various passages between the Windward Islands, into the Caribbean Sea. It sets toward the west, and then somewhat north of west, finally arriving off the Yucatan peninsula. From here, some of the water curves toward the right, flowing some distance off the shore of the Gulf of Mexico, and part of it curves more sharply toward the east and flows directly toward the north coast of Cuba. These two parts reunite in the Straits of Florida to form the most remarkable of all ocean currents, the **Gulf Stream.** Off the southeast coast of Florida this current is augmented by a current flowing along the northern coasts of Puerto Rico, Hispaniola, and Cuba. Another current flowing eastward of the Bahamas joins the stream north of these islands.

The Gulf Stream follows generally along the east coast of North America, flowing around Florida, northward and then northeastward toward Cape Hatteras, and then curving toward the east and becoming broader and slower. After passing the Grand Banks, it turns more toward the north and becomes a broad drift current flowing across the North Atlantic. That part in the Straits of Florida is sometimes called the **Florida current.**

A tremendous volume of water flows northward in the Gulf Stream. It can be distinguished by its deep indigo-blue color, which contrasts sharply with the dull green of the surrounding water. It is accompanied by frequent squalls. When the Gulf Stream encounters the cold water of the Labrador current, principally in the vicinity of the Grand Banks, there is little mixing of the waters. Instead, the junction is marked by a sharp change in temperature. The line or surface along which this occurs is called the **cold wall.** When the warm Gulf Stream water encounters cold air, evaporation is so rapid that the rising vapor may be visible as frost smoke. The stream carries large quantities of gulfweed from the tropics to higher latitudes.

Recent investigations have shown that the current itself is much narrower and faster than previously supposed, and considerably more variable in its position and speed. The maximum current off Florida ranges from about two to four knots. To the northward the speed is generally less, and decreases further after the current passes Cape

Hatteras. As the stream meanders and shifts position, eddies sometimes break off and continue as separate, circular flows until they dissipate. Boats in the Bermuda Race have been known to be within sight of each other and be carried in opposite directions by different parts of the same current. As the current shifts position, its extent does not always coincide with the area of warm, blue water. When the sea is relatively smooth, the edges of the current are marked by ripples.

Information is not yet available to permit prediction of the position and speed of the current at any future time, but it has been found that tidal forces apparently influence the current, which reaches its daily maximum speed about three hours after transit of the moon. The current generally is faster at the time of neap tides than at spring tides. When the moon is over the equator, the stream is narrower and faster than at maximum northerly or southerly declination. Variations in the trade winds also affect the current.

As the Gulf Stream continues eastward and northeastward beyond the Grand Banks, it gradually widens and decreases speed until it becomes a vast, slow-moving drift current known as the **North Atlantic current,** in the general vicinity of the prevailing westerlies. In the eastern part of the Atlantic it divides into the **northeast drift current** and the **southeast drift current.**

The northeast drift current continues in a generally northeasterly direction toward the Norwegian Sea. As it does so, it continues to widen and decrease speed. South of Iceland it branches to form the **Irminger current** and the **Norway current.** The Irminger current curves toward the north and northwest to join the East Greenland current southwest of Iceland. The Norway current continues in a northeasterly direction along the coast of Norway. Part of it, the **North Cape current,** rounds North Cape into the Barents Sea. The other part curves toward the north and becomes known as the **Spitzbergen current.** Before reaching Svalbard (Spitzbergen), it curves toward the west and joins the cold **east Greenland current** flowing southward in the Greenland Sea. As this current flows past Iceland, it is further augmented by the Irminger current.

Off Kap Farvel, at the southern tip of Greenland, the east Greenland current curves sharply to the northwest, following the coast line. As it does so, it becomes known as the **west Greenland current.** This current continues along the west coast of Greenland, through Davis Strait, and into Baffin Bay. Both east and west Greenland currents are sometimes known by the single name **Greenland current.**

In Baffin Bay the Greenland current follows generally the coast, curving westward off Kap York to form the southerly flowing **Labrador current.** This cold current flows southward off the coast of Baffin Island, through Davis Strait, along the coast of Labrador and Newfoundland, to the Grand Banks, carrying with it large quantities of ice. Here it encounters the warm water of the Gulf Stream, creating the "cold wall." Some of the cold water flows southward along the east coast of North America, inshore of the Gulf Stream, as far as Cape Hatteras. The remainder curves toward the east and flows along the northern edge of the North Atlantic and northeast drift currents, gradually merging with them.

The southeast drift current curves toward the east, southeast, and then south as it is deflected by the coast of Europe. It flows past the Bay of Biscay, toward southeastern Europe and the Canary Islands, where it continues as the **Canary current.** In the vicinity of the Cape Verde Islands, this current divides, part of it curving toward the west to help form the north equatorial current, and part of it curving toward the east to follow the coast of Africa into the Gulf of Guinea, where it is known as the **Guinea current.** This current is augmented by the equatorial countercurrent and, in summer, it is strengthened by monsoon winds. It flows in close proximity to the south equatorial current, but in the opposite direction. As it curves toward the south, still following the African coast, it merges with the south equatorial current.

The clockwise circulation of the North Atlantic leaves a large central area having no well-defined currents. This area is known as the **Sargasso Sea,** from the large quantities of sargasso or gulfweed encountered there.

That branch of the south equatorial current which curves toward the south off the east coast of South America follows the coast as the warm, highly-saline **Brazil current,** which in some respects resembles the Gulf Stream. Off Uruguay, it encounters the colder, less-salty Falkland current and the two curve toward the east to form the broad, slow-moving **South Atlantic current,** in the general vicinity of the prevailing westerlies. This current flows eastward to a point west of the Cape of Good Hope, where it curves northward to follow the west coast of Africa as the strong **Benguela current,** augmented somewhat by part of the Agulhas current flowing around the southern part of Africa from the Indian Ocean. As it continues northward, the current gradually widens and slows. At a point east of St. Helena Island it curves westward to continue as part of the south equatorial current, thus completing the counterclockwise circulation of the South Atlantic. The Benguela current is augmented somewhat by the **west wind drift,** a current which flows easterly around Antarctica. As the west wind drift flows past Cape Horn, that part in the immediate vicinity of the cape is called the **Cape Horn current.** This current rounds the cape and flows in a northerly and northeasterly direction along the coast of South America as the **Falkland current.**

Pacific Ocean currents follow the general pattern of those in the Atlantic. The **north equatorial current** flows westward in the general area of the northeast trades, and the **south equatorial current** follows a similar path in the region of the southeast trades. Between these two, the weaker **equatorial countercurrent** sets toward the east, just north of the equator.

After passing the Mariana Islands, the major part of the north equatorial current curves somewhat toward the northwest, past the Philippines and Formosa. Here it is deflected further toward the north, where it becomes known as the **Kuroshio,** and then toward the northeast past the Nansei Shoto and Japan, and on in a more easterly direction. Part of the Kuroshio, called the **Tsushima current,** flows through Tsushima Strait, between Japan and Korea, and the Sea of Japan, following generally the northwest coast of Japan. North of Japan it curves eastward and then southeastward to rejoin the main part of the Kuroshio. The limits and volume of the Kuroshio are influenced by the monsoons, being augmented during the season of southwesterly winds, and diminished when the northeasterly winds are prevalent.

The Kuroshio (Japanese for "Black Stream") is so named because of the dark color of its water. It is sometimes called the **Japan Stream.** In many respects it is similar to the Gulf Stream of the Atlantic. Like that current, it carries large quantities of warm tropical water to higher latitudes, and then curves toward the east as a major part of the general clockwise circulation in the northern hemisphere. As it does so, it widens and slows. A small part of it curves to the right to form a weak clockwise circulation west of the Hawaiian Islands. The major portion continues on between the Aleutians and the Hawaiian Islands, where it becomes known as the **North Pacific current.**

As this current approaches the North American continent, most of it is deflected toward the right to form a clockwise circulation between the west coast of North America and the Hawaiian Islands. This part of the current has become so broad that the circulation is generally weak. A small part near the coast, however, joins the southern branch of the Aleutian current, and flows southeastward as the **California current.** The average speed of this current is about 0.8 knot. It is strongest near land. Near the southern end of Baja (Lower) California, this current curves sharply to the west and broadens to form the major portion of the north equatorial current.

During the winter, a weak countercurrent flows northwestward along the west coast of North America from Southern California to Vancouver Island, inshore of the southeasterly flowing California current. This is called the **Davidson current.**

Off the west coast of Mexico, south of Baja California, the current flows southeastward, as a continuation of part of the California current, during the winter. During the summer, the current in this area is northwestward, as a continuation of the equatorial countercurrent, before it turns westward to help form the north equatorial current.

As in the Atlantic, there is in the Pacific a counterclockwise circulation to the north of the clockwise circulation. Cold water flowing southward through the western part of Bering Strait between Alaska and Siberia is joined by water circulating counterclockwise in the Bering Sea to form the **Oyashio.** As the current leaves the strait, it curves toward the right and flows southwesterly along the coast of Siberia and the Kuril Islands. This current brings quantities of sea ice, but no icebergs. When it encounters the Kuroshio, the Oyashio curves southward and then eastward, the greater portion joining the Kuroshio and North Pacific current. The northern portion continues eastward to join the curving Aleutian current.

As this current approaches the west coast of North America, west of Vancouver Island, part of it curves toward the right and is joined by water from the North Pacific current, to form the California current. The northern branch of the Aleutian current curves in a counterclockwise direction to form the **Alaska current,** which generally follows the coast of Canada and Alaska. When it arrives off the Aleutian Islands, it becomes known as the **Aleutian current.** Part of it flows along the southern side of these islands to about the 180th meridian, where it curves in a counterclockwise direction and becomes an easterly flowing current, being augmented by the northern part of the Oyashio. The other part of the Aleutian current flows through various openings between the Aleutian Islands, into the Bering Sea. Here it flows in a general counterclockwise direction, most of it finally joining the southerly flowing Oyashio, and a small part of it flowing northward through the eastern side of the Bearing Strait, into the Arctic Ocean.

The south equatorial current, extending in width between about 4° N latitude and 10° S, flows westward from South America to the western Pacific. After this current crosses the 180th meridian, the major part curves in a counterclockwise direction, entering the Coral Sea, and then curving more sharply toward the south along the east coast of Australia, where it is known as the **east Australia current.** In the Tasman Sea, northeast of Tasmania, it is augmented by water from the west wind drift, flowing eastward south of Australia. It curves toward the southeast and then the east, gradually merging with the easterly flowing west wind drift, a broad, slow-moving current that circles Antarctica.

Near the southern extremity of South America, most of this current flows eastward into the Atlantic, but part of it curves toward the left and flows generally northward along the west coast of South America as the **Peru current.** Occasionally a set directly toward land is encountered. At about Cabo Blanco, where the coast falls away to the right, the current curves toward the left, past the Galapagos Islands, where it takes a westerly set and constitutes the major portion of the south equatorial current, thus completing the counterclockwise circulation of the South Pacific.

During the northern hemisphere summer, a weak northern branch of the south equatorial current, known as the **Rossel current,** continues on toward the west and northwest along both the southern and northeastern coasts of New Guinea. The southern part flows through Torres Strait, between New Guinea and Australia, into the Arafura Sea. Here, it gradually loses its identity, part of it flowing on toward the west as part of the south equatorial current of the Indian Ocean, and part of it following the coast of Australia and finally joining the easterly flowing west wind drift. The northern part of the Rossel current curves in a clockwise direction to help form the Pacific equatorial countercurrent. During the northern hemisphere winter, the Rossel current is replaced by an easterly flowing current from the Indian Ocean.

Indian Ocean currents follow generally the pattern of the Atlantic and Pacific, but with differences caused pincipally by the monsoons and the more limited extent of water in the northern hemisphere. During the northern hemisphere winter, the **north equatorial current** and **south equatorial current** flow toward the west, with the weaker, easterly flowing **equatorial countercurrent** flowing between them, as in the Atlantic and Pacific (but somewhat south of the equator). But during the northern hemisphere summer, both the north equatorial current and the equatorial countercurrent are replaced by the **monsoon current,** which flows eastward and southeastward across the Arabian Sea and the Bay of Bengal. Near Sumatra, this current curves in a clockwise direction and flows westward, augmenting the south equatorial current and setting up a clockwise circulation in the northern part of the Indian Ocean.

As the south equatorial current approaches the coast of Africa, it curves toward the southwest, part of it flowing through the Mozambique Channel between Madagascar and the mainland, and part flowing along the east coast of Madagascar. At the southern end of this island the two join to form the strong **Agulhas current,** which is analogous to the Gulf Stream.

A small part of the Agulhas current rounds the southern end of Africa and helps form the Benguela current. The major portion, however, curves sharply southward and then eastward to join the west wind drift. This junction is often marked by a broken and confused sea. During the northern hemisphere winter the northern part of this current curves in a counterclockwise direction to form the **West Australia current,** which flows northward along the west coast of Australia. As it passes Northwest Cape, it curves northwestward to help form the south equatorial current. During the northern hemisphere summer, the West Australia current is replaced by a weak current flowing around the western part of Australia as an extension of the southern branch of the Rossel current.

Polar Currents—The waters of the North Atlantic enter the Arctic Ocean between Norway and Svalbard. The currents flow easterly north of Siberia to the region of the Novosibirskiye Ostrova, where they turn northerly across the north pole and continue down the Greenland coast to form the east Greenland current. On the American side of the arctic basin, there is a weak, continuous clockwise flow centered in the vicinity of 80° N, 150° W. A current north through Bering Strait along the American coast is balanced by an outward southerly flow along the Siberian coast, which eventually becomes part of the Oyashio. Each of the main islands or island groups in the arctic, as far as is known, seems to have a clockwise nearshore circulation around it. The Barents Sea, Kara Sea, and Laptev Sea each have a weak counterclockwise circulation. A similar but weaker counterclockwise current system appears to exist in the East Siberian Sea.

In the antarctic, the circulation is generally from west to east in a broad, slow-moving current extending completely around Antarctica. This is called the **west wind drift,** although it is formed partly by the strong westerly wind in this area and partly by density differences. This current is augmented by the Brazil and Falkland currents in the Atlantic, the east Australia current in the Pacific, and the Agulhas current in the Indian Ocean. In return, part of it curves northward to form the Cape Horn, Falkland, and most of the Benguela currents in the Atlantic, the Peru current in the Pacific, and west Australia current in the Indian Ocean.

Ocean Currents and Climate—Many of the ocean currents exert a marked influence upon the climate of the coastal regions along which they flow. Thus, warm water from the Gulf Stream, continuing as the North Atlantic, northeast drift, and Irminger currents, arrives off the southwest coast of Iceland, warming it to the extent that Reykjavík has a higher average winter temperature than New York City, far to the south. Great Britain and Labrador are at about the same latitude, but the climate of Great Britain is much milder because of the difference of temperature of currents. The West Coast of the United States is cooled in the summer by the California current, and warmed in the winter by the Davidson current. As a result of this condition, partly, the range of monthly average temperature is comparatively small.

Currents exercise other influences besides those on temperature. The pressure pattern is affected materially, as air over a cold current contracts as it is cooled, and that over a warm current expands. As air cools above a cold ocean current, fog is likely to form. Frost smoke is most prevalent over a warm current which flows into a colder region. Evaporation is greater from warm water than from cold water.

In these and other ways, the climate of the earth is closely associated with the ocean currents, although other factors, such as topography and prevailing winds, are also important.

=5=

Tides and Tidal Currents

The tidal phenomenon is the periodic motion of the waters of the sea due to differences in the attractive forces of various celestial bodies, principally the moon and sun, upon different parts of the rotating earth. It can be either a help or hindrance to the mariner—the water's rise and fall may at certain times provide enough depth to clear a bar and at others may prevent him from entering or leaving a harbor. The flow of the current may help his progress or hinder it, may set him toward dangers or away from them. By understanding this phenomenon and by making intelligent use of predictions published in tide and tidal current tables and of descriptions in sailing directions, the mariner can set his course and schedule his passage to make the tide serve him, or at least to avoid its dangers.

Tide and Current—In its rise and fall, the tide is accompanied by a periodic horizontal movement of the water called **tidal current.** The two movements, tide and tidal current, are intimately related, forming parts of the same phenomenon brought about by the tide-producing forces of the sun and moon, principally.

It is necessary, however, to distinguish clearly between tide and tidal current, for the relation between them is not a simple one nor is it everywhere the same. For the sake of clearness and to avoid misunderstanding, it is desirable that the mariner adopt the technical usage: **tide** for the vertical rise and fall of the water, and **current** for the horizontal flow. The tide rises and falls, the tidal current floods and ebbs.

Cause Tides result from differences in the gravitational attraction of various celestial bodies, principally the moon and sun, upon different parts of the rotating earth. The gravity of the earth acts approximately toward the earth's center, and tends to hold the earth in the shape of a sphere. But the moon and sun provide disturbing, or tide-producing, forces. Consider the earth and moon. The moon appears to revolve about the earth, but actually the moon and earth revolve about their common center of mass. They are held together by gravitational attraction and kept apart by an equal and opposite centrifugal force. In this earth-moon system, the tide-producing force on the earth's hemisphere nearer the moon is in the direction of the moon's attraction, or toward the moon. On the hemisphere opposite the moon the tide-producing force is in the direction of the centrifugal force, or away from the moon.

At the sublunar point, and its antipode, the moon's attractive force is vertical, in the opposite direction to gravity. Along the great circle midway between these points, the force is horizontal, parallel to the earth's surface. At any other point, the moon's tide-producing force can be resolved into horizontal and vertical components. Both are very small compared to the earth's gravity. Since the horizontal component is not operating against gravity and can draw particles of water over the surface of the earth, it is the more effective in generating tides.

The tide-producing forces, then, tend to create high tides on the sides of the earth nearest to and farthest from the moon, with a low tide belt between them. As the earth rotates, a point on earth passes through two high and two low areas each day if the moon is over the equator as shown in figure 5–1 A. When the moon is north or south of the equator, the force pattern is as shown in figure 5–1 B, and a point on the equator passes through two equal highs, but a point in higher latitudes passes through two unequal highs or only one high. Thus, due to changes in the moon's declination, there is introduced a diurnal inequality in the pattern of the tidal forces at a particular place. There are similar forces due to the sun, and the total tide producing force is the resultant of the two. Minute tidal effects are caused by other celestial bodies.

The mathematician develops his formulas by considering the difference in attraction between a point on the earth's surface and a point at the earth's center. In accordance with Newton's law, gravitational attraction of an astronomical body varies directly as its mass and inversely as the *square* of its distance. But the tide-producing (differential) force varies directly as the mass and inversely as the *cube* of the distance. As a consequence, only the moon and sun produce any appreciable tidal effect upon the earth. Further, although the moon's mass is but a fraction of the sun's, dividing such masses by the cube of their respective distances—$(238,862)^3$ statute miles

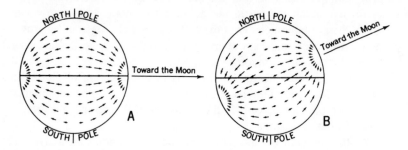

Figure 5–1 *Tide-producing forces. The arrows represent the magnitude and direction of the horizontal component of the tide-producing force on the earth's surface. A. When the moon is in the plane of the equator, the forces are equal in magnitude at the two points on the same parallel of latitude and 180° apart in longitude. B. When the moon is at north (or south) declination, the forces are unequal at such points and tend to cause an inequality in the two high waters and the two low waters of a day.*

and $(92,900,000)^3$ statute miles, respectively—reduces the sun's tide-producing force to only 0.46 that of the moon. It is because of this that the timing of the tides is identified so closely with the motions of the moon.

Though the tide-producing forces are distributed over the earth in a regular manner, the sizes and shapes of the ocean basins and the interference of the land masses prevent the tides of the oceans from assuming a simple, regular pattern. The way in which the waters in different parts of the oceans, as well as in the smaller waterways, respond to these known regular forces is dependent in large part upon the size, depth, and configuration of the basin or waterway.

General Features—Tide is the periodic rise and fall of the water accompanying the tidal phenomenon. At most places it occurs twice daily. The tide rises until it reaches a maximum height, called **high tide** or **high water,** and then falls to a minimum level called **low tide** or **low water.**

The rate of rise and fall is not uniform. From low water, the tide begins to rise slowly at first but at an increasing rate until it is about halfway to high water. The rate of rise then decreases until high water is reached and the rise ceases. The falling tide behaves in a similar manner. The period at high or low water during which there is no sensible change of level is called **stand.** The difference in height between consecutive high and low waters is the **range.**

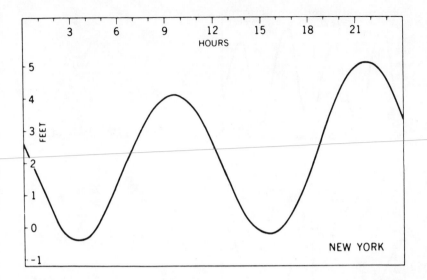

Figure 5–2 *The rise and fall of the tide at New York, shown graphically.*

Types of Tide—A body of water has a natural period of oscillation that is dependent upon its dimensions. None of the oceans appears to be a single oscillating body, but rather each one is made up of a number of oscillating basins. As such basins are acted upon by the tide-producing forces, some respond more readily to daily or diurnal forces, others to semidiurnal forces, and others almost equally to both. Hence, tides at a place are classified as one of three types— **semidiurnal, diurnal,** or **mixed**—according to the characteristics of the tidal pattern occurring at the place.

In the **semidiurnal** type of tide, there are two high and two low waters each tidal day, with relatively small inequality in the high and low water heights. Tides on the Atlantic coast of the United States are representative of the semidiurnal type, which is illustrated in figure 5–3 by a tide curve chart for Boston Harbor.

In the **diurnal** type of tide, only a single high and single low water occur each tidal day. Tides of the diurnal type occur along the northern shore of the Gulf of Mexico, in the Java Sea, the Gulf of Tonkin, and in a few other localities. The tide curve for Pakhoi, China, illustrated in figure 5–4 is an example of the diurnal type.

In the **mixed** type of tide, the diurnal and semidiurnal oscillations

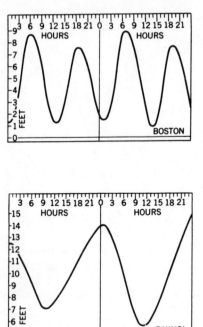

Figure 5–3 *Semidiurnal type of tide.*

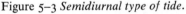

Figure 5–4 *Diurnal type of tide.*

are both important factors and the tide is characterized by a large inequality in the high water heights, low water heights, or in both. There
are usually two high and two low waters each day, but occasionally
the tide may become diurnal. Such tides are prevalent along the
Pacific coast of the United States and in many other parts of the
world. Examples of mixed types of tide are shown in figure 5–5. At
Los Angeles, it is typical that the inequalities in the high and low
waters are about the same. At Seattle the greater inequalities are
typically in the low waters, while at Honolulu it is the high waters that
have the greater inequalities.

Solar Tide—The natural period of oscillation of a body of water
may accentuate either the solar or the lunar tidal oscillations. Though
it is a general rule that the tides follow the moon, the relative importance of the solar effect varies in different areas. There are a few
places, primarily in the South Pacific and the Indonesian areas, where
the solar oscillation is the more important, and at those places the
high and low waters occur at about the same time each day. At Port
Adelaide, Australia, figure 5–6, the solar and lunar semidiurnal oscil-

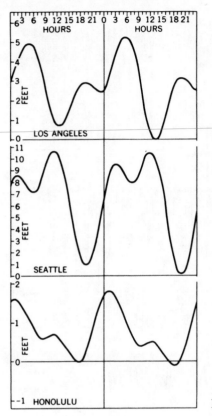

Figure 5–5 *Mixed types of tide.*

lations are equal and nullify one another at neaps, which will be discussed later.

Special Effects—As a progressive wave enters shallow water, its speed is decreased. Since the trough is shallower than the crest, its retardation is greater, resulting in a steepening of the wave front. Therefore, in many rivers, the duration of rise is considerably less than the duration of fall. In a few estuaries, the advance of the low water trough is so much retarded that the crest of the rising tide overtakes the low, and advances upstream as a churning, foaming wall of water called a **bore.** Bores that are large and dangerous at times of large tidal ranges may be mere ripples at those times of the month when the range is small. The tide tables indicate where bores occur.

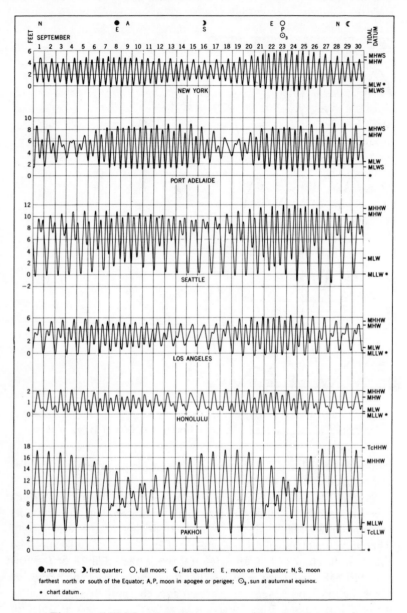

Figure 5–6 *Tidal variations at various places during a month.*

Other special features are the **double low water** (as at Hoek Van Holland) and the **double high water** (as at Southampton, England). At such places there is often a slight fall or rise in the middle of the high or low water period. The practical effect is to create a longer period of stand at high or low tide. The tide tables direct attention to these and other peculiarities where they occur.

Variations in Range—Though the tide at a particular place can be classified as to type, it exhibits many variations during the month. The range of the tide varies in accordance with the intensity of the tide-producing force, though there may be a lag of a day or two **(age of tide)** between a particular astronomic cause and the tidal effect.

Thus, when the moon is at the point in its orbit nearest the earth (at *perigee*), the lunar semidiurnal range is increased and **perigean** tides occur; when the moon is farthest from the earth (at *apogee*), the smaller **apogean** tides occur. When the moon and sun are in line and pulling together, as at new and full moon, **spring** tides occur (the term *spring* has nothing to do with the season of year); when the moon and sun oppose each other, as at the quadratures, the smaller **neap** tides occur.

When certain of these phenomena coincide, the great **perigean spring** tides, the small **apogean neap** tides, etc., occur.

These are variations in the semidiurnal portion of the tide. Variations in the diurnal portion occur as the moon and sun change declination. When the moon is at its maximum semi-monthly declination (either north or south), **tropic** tides occur in which the diurnal effect is at a maximum; when it crosses the equator, the diurnal effect is at a minimum and **equatorial** tides occur.

It should be noted that when the range of tide is increased, as at spring tides, there is more water available only at *high* tide; at *low* tide there is less, for the high waters rise higher and the low waters fall lower at these times. There is more water at neap low water than at spring low water. With tropic tides, there is usually more depth at one low water during the day than at the other. While it is desirable to know the meanings of these terms, the best way of determining the height of the tide at any place and time is to examine the tide predictions for the place as given in the tide tables. Figure 5–7 shows variations in the ranges and heights of tides in a locality where the water level always exceeds the charted depth.

Tidal Cycles—Tidal oscillations go through a number of cycles. The shortest cycle, completed in about 12 hours and 25 minutes for a semidiurnal tide, extends from any phase of the tide to the next re-

TIDES AND TIDAL CURRENTS

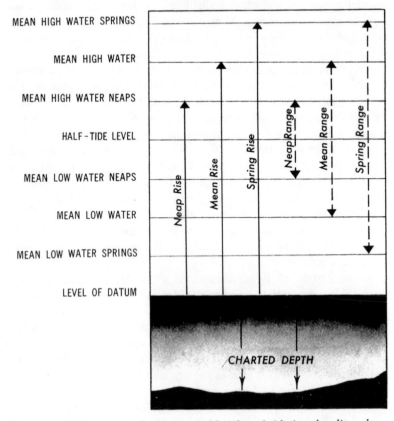

MEAN HIGH WATER SPRINGS

MEAN HIGH WATER

MEAN HIGH WATER NEAPS

HALF-TIDE LEVEL

MEAN LOW WATER NEAPS

MEAN LOW WATER

MEAN LOW WATER SPRINGS

LEVEL OF DATUM

Neap Rise *Mean Rise* *Spring Rise* *Neap Range* *Mean Range* *Spring Range*

CHARTED DEPTH

Figure 5–7 *Variations in the ranges and heights of tide in a locality where the water level always exceeds the charted depth.*

currence of the same phase. During a **lunar day** (averaging 24 hours and 50 minutes) there are two highs and two lows (two of the shorter cycles) for a semidiurnal tide. The effect of the phase variation is completed in about two weeks as the moon varies from new to full or full to new. The effect of the moon's declination is also repeated about each two weeks. The cycle involving the moon's distance requires approximately a **lunar month** (a synodical month of about 29½ days). The sun's declination and distance cycles are respectively a half year and a year in length. An important lunar cycle, called the **nodal period,** is 18.6 years (usually expressed in round figures as 19 years).

For a tidal value, particularly a range, to be considered a true mean, it must be either based upon observations extended over this period of time or adjusted to take account of variations known to occur during the cycle.

Time of Tide—Since the lunar tide-producing force has the greater effect in producing tides at most places, the tides "follow the moon." Because of the rotation of the earth, high water lags behind meridian passage (upper and lower) of the moon. The **tidal day,** which is also the **lunar day,** is the time between consecutive transits of the moon, or 24 hours and 50 minutes on the average. Where the tide is largely semidiurnal in type, the **lunitidal interval**—the interval between the moon's meridian transit and a particular phase of tide—is fairly constant throughout the month, varying somewhat with the tidal cycles. There are many places, however, where solar or diurnal oscillations are effective in upsetting this relationship, and the newer editions of charts of many countries now omit intervals because of the tendency to use them for prediction even though accurate predictions are available in tide tables. However, the lunitidal interval may be encountered. The interval generally given is the average elapsed time from the meridian transit (upper or lower) of the moon until the next high tide. This may be called **mean high water lunitidal interval** or **establishment of the port.** The **high water full and change (HWF&C)** or **vulgar establishment,** sometimes given, is the average interval on days of full or new moon, and approximates the mean high water lunitidal interval.

In the ocean, the tide may be of the nature of a progressive wave with the crest moving forward, a stationary or standing wave which oscillates in a seesaw fashion, or a combination of the two. Consequently, caution should be used in inferring the time of tide at a place from tidal data for nearby places. In a river or estuary, the tide enters from the sea and is usually sent upstream as a progressive wave, so that the tide occurs progressively later at various places upstream.

Tidal Datums—A **tidal datum** is a level from which heights and depths are measured. There are a number of such levels of reference that are important to the mariner.

The most important level of reference to the mariner is the datum of soundings on charts. Since the tide rises and falls continually while soundings are being taken during a hydrographic survey, the tide should be observed during the survey so that soundings taken at all stages of the tide can be reduced to a common datum. Soundings on charts show depths below a selected low water datum (occasionally mean sea level), and tide predictions in tide tables show heights above the same level. The depth of water available at any time is obtained by adding the height of the tide at the time in question to the charted depth, or by subtracting the predicted height if it is negative.

By international agreement, the level used as chart datum should be just low enough so that low waters do not go far below it. At most places, however, the level used is one determined from a mean of a number of low waters (usually over a 19-year period); therefore some low waters can be expected to fall below it. The following are some of the datums in general use.

The highest low water datum in considerable use is **mean low water (MLW),** which is the average height of all low waters at a place. About half of the low waters fall below it. **Mean low water springs (MLWS),** usually shortened to **low water springs,** is the average level of the low waters that occur at the times of spring tides. **Mean lower low water (MLLW)** is the average height of the lower low waters at a place. **Tropic lower low water (TcLLW)** is the average height of the lower low waters (or of the single daily low waters if the tide becomes diurnal) that occur when the moon is near maximum declination and the diurnal effect is most pronounced. This datum is not in common use as a tidal reference. **Indian spring low water (ISLW)** sometimes called **Indian tide plane** or **harmonic tide plane,** is a low datum that includes the spring effect of the semidiurnal portion of the tide and the tropic effect of the diurnal portion. It is about the level of lower low water of mixed tides at the time that the moon's maximum declination coincides with the time of new or full moon. **Mean lower low water springs** is the average level of the lower of the two low waters on the days of spring tides. Some still lower datums used on charts are determined from tide observations and some are determined arbitrarily and later referred to the tide. Most of them fall close to one or the other of the following two datums. **Lowest normal low water** is a datum that approximates the average height of monthly lowest low waters, discarding any tides disturbed by storms. **Lowest low water** is an extremely low datum. It conforms generally to the lowest tide observed, or even somewhat lower. Once a tidal datum is established, it is generally retained for an indefinite period, even though it might differ slightly from a better determination from later observations. When this occurs, the established datum may be called **low water datum, lower low water datum,** etc.

In some areas where there is little or no tide, such as the Baltic Sea, **mean sea level (MSL)** is used as chart datum. This is the average height of the surface of the sea for all stages of the tide over a 19-year period. This may differ slightly from **half-tide level,** which is the level midway between mean high water and mean low water.

Inconsistencies of terminology are found among charts of different countries and between charts issued at different times. For example, the spring effect as defined here is a feature of only the semidiurnal tide, yet it is sometimes used synonymously with tropic effect to refer to times of increased range of a diurnal tide. Such inconsistencies are being reduced through increased international cooperation.

Large-scale charts usually specify the datum of soundings and may contain a tide note giving mean heights of the tide at one or more places on the chart. These heights are intended merely as a rough guide to the change in depth to be expected under the specified conditions. They should not be used for the prediction of heights on any particular day. Such predictions should be obtained from *tide tables*.

High Water Datums—Heights of land features are usually referred on nautical charts to a high water datum. The one used on charts of the United States, its territories, and possessions, and widely used elsewhere, is **mean high water (MHW),** which is the average height of all high waters over a 19-year period. Any other high water datum in use on charts is likely to be higher than this. Other high water datums are **mean high water springs (MHWS),** which is the average level of the high waters that occur at the time of spring tides; **mean higher high water (MHHW),** which is the average height of the higher high waters of each day; and **tropic higher high water (TcHHW),** which is the average height of the higher high waters (or the single daily high waters if the tide becomes diurnal) that occur when the moon is near maximum declination and the diurnal effect is most pronounced. A reference merely to "high water" leaves some doubt as to the specific level referred to, for the height of high water varies from day to day. Where the range is large, the variation during a two-week period may be considerable.

Observations and Predictions—Since the tide at different places responds differently to the tide-producing forces, the nature of the tide at any place can be determined most accurately by actual observation. The predictions in tide tables and the tidal data on nautical charts are based upon such observations.

Tides are usually observed by means of a continuously recording gage. A year of observations is the minimum length desirable for determining the **harmonic constants** used in prediction. For establishing mean sea level and the long-time changes in the relative elevations of land and sea, as well as for other special uses, observations have been made over periods of 20, 30, and even 50 years at important locations. Observations for a month or less will establish the *type* of tide and suffice for comparison with a longer series of a similar type to determine tidal differences and constants.

Sailing directions and coast pilots issued by maritime nations include general descriptions of current behavior in various localities throughout the world.

Tidal Current Charts—A number of important harbors and waterways are covered by sets of tidal current charts showing graphically the hourly current movement.

=6=

Ice in the Sea

Ice and the Navigator—The perpetually frozen Arctic Ocean and the solid sheet of ice beneath which Antarctica is buried offer evidence that the earth has not yet completely emerged from its most recent Ice Age. Each winter this polar ice increases and spreads toward more temperate latitudes, and each summer it contracts again as part of the ice melts. Some of the fragments are carried by ocean currents into shipping lanes, forming a major hazard to shipping. There is evidence to indicate that the polar regions are becoming warmer. Nearly all glaciers are receding; the ice shelves off northern Canada and Greenland are breaking up; shipping off the Siberian coast has become possible; cod are found ever farther north along the Greenland coast.

Ice is of direct concern to the navigator because it restricts and sometimes controls his movements, it affects his dead reckoning by forcing frequent and sometimes inaccurately determined changes of course and speed, it affects his piloting by altering the appearance or obliterating the features of landmarks and by rendering difficult the establishment and maintenance of aids to navigation, it affects his electronic navigation by its effect upon propagation of radio waves and the changes it produces both in surface features and radar returns from such features, it affects his celestial navigation by altering the refraction and obscuring his horizon and celestial bodies either directly or by the weather it influences, and it affects his charts by introducing various difficulties to the hydrographic surveyor.

Because of his direct concern with ice, the prospective polar navigator will do well to acquaint himself with its nature and extent in the area he expects to navigate. To this end he should consult the sailing directions for the area, and whatever other literature may be available to him, including reports of previous operations in the same area.

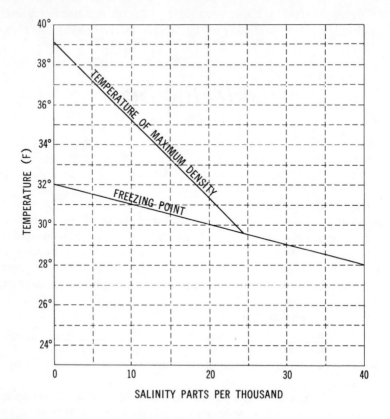

Figure 6–1 *Relationship between temperature of maximum density and freezing point for water of varying salinity.*

Formation of Ice—As it cools, water contracts until the temperature of maximum density is reached. Further cooling results in expansion. The maximum density of fresh water occurs at a temperature of 39°.2 F, and freezing takes place at 32° F. The addition of salt lowers both the temperature of maximum density and, to a lesser extent, that of freezing. The relationships are shown in figure 6–1. The two lines meet at a salinity of 24.7 parts per thousand, at which maximum density occurs at the freezing temperature of 29°.61 F. At this and greater salinities, the density increases right down to the freezing point. At a salinity of 35 parts per thousand, the approximate average for the oceans, the freezing point is 28°.6 F.

Generally, ice forms first at the water surface. As it does, most of the dissolved solids remain in the water, beneath the ice, increasing the density of the water there. This lowers the freezing point, thus tending to retard the freezing process. It is further retarded by the fact that ice is a poor conductor of heat and therefore serves as an insulator to protect the water from colder air above.

In shoal water and streams, particularly where motion is sufficient to cause thorough mixing, the freezing temperature may extend from the surface to the bottom. When this occurs, ice crystals may form at any depth. Because of their decreased density, they tend to rise to the surface, unless they form at the bottom and attach themselves there. This **bottom ice,** sometimes called **anchor ice,** continues to grow as additional ice freezes to that already formed.

Ice may also be formed by the compacting of fallen snow, or by the freezing of a mixture of snow and sea water.

Land ice is formed on land by the freezing of fresh water or the compacting of snow as layer upon layer adds to the pressure on that beneath. As snow becomes hardened by wind, temperature, and pressure, it reaches an intermediate stage when it is known as **névé** (nā′ vā′).

Under great pressure ice becomes slightly plastic and is forced outward and downward along an inclined surface. If a large area is relatively flat, as on the antarctic plateau, or if the outward flow is obstructed, as on Greenland, an **ice cap** forms and remains winter and summer, in some places reaching depths of several thousand feet. Where ravines or mountain passes permit flow of the ice, a **glacier** is formed. This is a slow-moving river of ice that flows to lower levels, exhibiting many of the characteristics of rivers of water. The flow may be more than 100 feet per day, but is generally much less. When a glacier reaches a comparatively level area, it spreads out. When a glacier flows into the sea, the buoyant force of the water breaks off pieces from time to time, and these float away as **icebergs.**

An iceberg seldom melts uniformly because of lack of uniformity in the ice itself, differences in the temperature above and below the water line, exposure of one side to the sun, strains, cracks, mechanical erosion, etc. The inclusion of rocks, silt, and other foreign matter further accentuates the differences. As a result, changes in equilibrium take place, which may cause the berg to tilt or capsize. Parts of it may break off or **calve,** forming separate, smaller bergs. A small berg about the size of a house is called a **bergy bit,** and one still smaller but large enough to inflict serious damage to a vessel is called a **growler** because of the noise it sometimes makes as it bobs up and down in the sea. Bergy bits and growlers are usually pieces calved from icebergs, but they may be formed by consolidation of sea ice or by the melting of an iceberg. The principal danger from icebergs is their tendency to break or shift position, and possible underwater extensions, called **rams.**

Sea ice forms by the freezing of sea water. The first indication is a greasy or oily appearance of the surface, with a peculiar gray or leaden tint. The small individual particles of ice, called **spicules,** then become visible. As the number increases, the mixture of water and ice is soupy or mushy, having about the consistency of wet snow. At this stage it is called **slush.** The height of waves is noticeably reduced. As the individual particles freeze together, a thin layer of highly plastic ice forms. This bends easily and moves up and down with the waves. A layer of two inches of fresh-water ice is brittle but strong enough to support the weight of a heavy man. In contrast, the same thickness of newly formed sea ice will support not more than about ten percent of this weight, although its strength varies with the temperature at which it is formed, very cold ice supporting a greater weight than warmer ice. When snow falls into sea water which is near its freezing point, but colder than the melting point of snow, it does not melt, but floats on the surface, drifting with the wind into beds which may become several feet thick. If the temperature drops below the freezing point of the sea water, the mixture of snow and water freezes quickly into a soft ice similar to that formed when snow is not present. As it ages, sea ice becomes harder and more brittle.

Close to land the ice may be attached to the shore as an **ice foot.** The width of this **fast ice** varies considerably, but in an area with many irregularities in the coast line, especially if there are offshore islands or shoals, and relatively shallow water, it may extend for several miles to seaward. Although the width generally varies from two to 20 miles, a maximum of about 270 miles has been observed in the vicinity of Novosibirskiye Ostrova (New Siberian Islands). On an exposed, abrupt coast bordered by deep water there may be no ice foot at all.

In a bay or other sheltered area, ice formed on the surface of the sea, often augmented by snow and land ice, may build up a shelf which remains attached to the land for many years. In the Ross Sea in Antarctica this **shelf ice** attains a thickness of 500 to 1,000 feet. At the outer edge, large pieces eventually break away, forming **tabular icebergs,** with dimensions measured in *miles*. In 1854 and 1855

Figure 6–2 *A tabular iceberg.*

several ships in the South Atlantic reported a crescent-shaped iceberg with one horn 40 miles long, the other 60 miles long, and with an embayment 40 miles wide between the tips. In 1927 a berg 100 miles long, 100 miles wide, and 130 feet high above water was reported. The largest iceberg ever reported was sighted in 1956 by the USS *Glacier*, a U. S. Navy icebreaker, about 150 miles west of Scott Island. This berg was 60 miles wide and 208 miles long, more than twice the size of Connecticut.

Icebergs ten miles or more in length have been seen on many occasions in the antarctic. In contrast, the largest iceberg reported in the northern hemisphere was seven miles long and three and a half miles wide. This berg was sighted off Baffin Island in 1882. In 1928 an iceberg four miles long was reported seen in the North Atlantic. The expression "tabular iceberg" is not applied to northern hemisphere bergs, but similar formations there are called **ice islands.** These are believed to originate when shelf ice breaks up north of Canada and Greenland. Most of them remain in the Arctic Ocean and have not been encountered by ships, although the large icebergs sighted in 1882 and 1928 were possibly ice islands. For several years the United States maintained a weather station on one of the arctic ice islands.

Sea ice is exposed to several forces, including currents, wave motion, tides, wind, and temperature differences. In its early stages, its plasticity permits it to conform readily to virtually any shape required by the forces acting upon it. As it becomes older, thicker, and more brittle, exposed sea ice cracks and breaks under the strain. Under the influence of wind and current, the broken pieces may shift position relative to pieces around them.

A single piece of relatively flat sea ice is called an **ice cake.** When ice is formed in the presence of considerable wave motion, circular

Figure 6–3 *Pancake ice, with an iceberg in the background.*

cakes several feet in diameter are formed, rather than a single large sheet. These circular cakes are called **pancakes,** and a collection of pancakes is called **pancake ice,** illustrated in figure 6–3. Wave motion may cause the pancakes to break into smaller pieces. With continued freezing, individual pieces unite into **floes,** and floes into **ice fields** which extend over many miles.

When one floe encounters another, or the shore, the individual pieces may be forced closer together into a thickly compacted mass. If the force is sufficient, and the ice is sufficiently plastic, **bending** takes place, or **tenting** if the contacting edges of individual cakes force each other to rise above their surroundings. More frequently, **rafting** occurs as one cake overrides another. Sea ice having any readily observed roughness of the surface is called **pressure ice.** A line of ice piled haphazardly along the edge of two floes which have collided is called a **pressure ridge.** Pressure ice with numerous mounds or hillocks which have become somewhat rounded and smooth by weathering or the accumulation of snow is called **hummocked ice,** each mound being called a **hummock.**

The motion of adjacent floes is seldom equal. The rougher the surface, the greater the effect of wind, since each piece extending above the surface acts as a sail. Some floes are in rotary motion as they tend to trim themselves into the wind. Since ridges extend below as well as above the surface, the deeper ones are influenced more by deep-water currents. When a strong wind blows in the same direction for a considerable period, each floe exerts pressure on the next one, and as the distance increases, the pressure becomes tremendous. Near land the result is an almost unbelievably chaotic piling of ice. Individual ridges near the shore may extend as much as 60 or 70 feet above surrounding ice and have a total thickness of 150 to 200 feet in extreme cases. Far from land, the height and thickness seldom exceed half these figures.

The continual motion of various floes results in separation as well as consolidation. A long, narrow, jagged **crack** may appear and widen enough to permit passage of a ship, when it is called a **lead** (lēd). In winter, a thin coating of newly formed ice usually covers the water, but in summer the water remains ice-free until a shift in the movement forces the two sides together again. Before this occurs, lateral motion usually takes place between the floes, so that they no longer fit, and unless the pressure is extreme, numerous patches of open water remain. A large one is called a **polynya.**

A large mass of sea ice, consisting of various floes, pressure ridges, and openings, is called a **pack.** In the arctic the main pack extends over the entire Arctic Ocean and for a varying distance outward from it, the limits receding considerably during summer. Each year a large portion of the ice from the Arctic Ocean moves outward between Greenland and Norway, into the North Atlantic, and is replaced by new ice. Relatively little of the pack ice is more than ten years old. The **ice pole,** the approximate center of the arctic pack, is at latitude 83°.5 N, longitude 160° W, north of western Alaska and about 390 miles from the north pole. In the antarctic the pack exists as a relatively narrow strip between the continent of Antarctica and the notoriously stormy seas which hasten the pack's destruction.

The alternate melting and refreezing of the surface of the pack, producing **weathered ice,** combined with the various motions to which the pack is subjected, result in widely varying conditions within the pack itself. The extent to which it can be penetrated by a ship varies from place to place and with changing weather conditions. In some areas the limit of navigable water is abrupt and complete, as at the edge of shelf ice. Such ice is called a **barrier.**

Thickness of Sea Ice—The seasonal thickness of fast ice in two harbors of the northern hemisphere is shown in figure 6–4, at the latitudes indicated. Pack ice in these latitudes undergoes a similar change. As ice thickens, it provides increased insulation to protect the

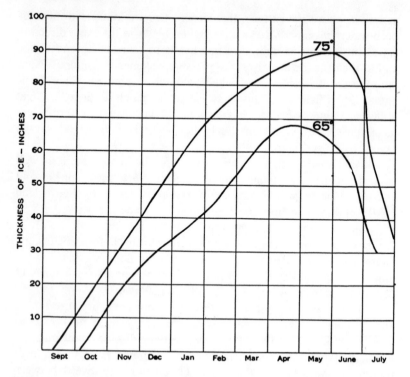

Figure 6–4 *Thickness of ice in two typical sheltered harbors in the northern hemisphere, at the latitudes indicated.*

sea water beneath from the colder air above, and the rate of freezing decreases. Sea ice rarely exceeds six feet in thickness during its first year. In a coastal area where the melting rate is less than the freezing rate, the thickness increases during succeeding winters, being augmented by compacted and frozen snow, until a maximum thickness of about 12 to 15 feet may eventually be reached. These values refer to single, unbroken pieces of floating ice. Shelf ice and pressure ice may be much thicker, as indicated previously.

During the summer, the sea ice insulates the sea water from warmer air above, so that melting is confined almost entirely to the upper portion. As the fresher melt water runs off into the sea, it tends to float on top of the heavier and colder salt water of the ocean. The temperature of the sea water may be lower than the freezing point of the fresher melt water, resulting in some refreezing as the melt water runs under the ice.

Salinity of Sea Ice—Sea ice forms first as salt-free crystals near the surface of the sea. As the process continues, these crystals are joined together and, as they do so, small quantities of brine are trapped within the ice. On the average, new ice six inches thick contains five to ten parts of salt per thousand. With lower temperature, freezing takes place faster. With faster freezing, a greater amount of salt is trapped in the ice.

Depending upon the temperature, the trapped brine may either freeze or remain liquid, but because its density is greater than that of the pure ice, it tends to settle down through the pure ice. As it does so, the ice gradually freshens, becoming clearer, stronger, and more brittle. At an age of one year sea ice is sufficiently fresh that its melt water, if found in **puddles** of sufficient size, and not contaminated by spray from the sea, can be used to replenish the fresh water supply of a ship. However, ponds of sufficient size to water ships are seldom found except in ice of great age, and then much of the melt water is from snow which has accumulated on the surface of the ice. When sea ice reaches an age of about two years, virtually all of the salt has been eliminated. Icebergs contain no salt, and uncontaminated melt water obtained from them is fresh.

The settling out of the brine gives sea ice a honeycomb structure which greatly hastens its disintegration when the temperature rises above freezing. In this state, when it is called **rotten ice,** much more surface is exposed to warm air and water, and the rate of melting is increased. In a day's time, a floe of apparently solid ice several inches thick may disappear completely.

Density of Ice—The density of fresh-water ice at its freezing point is 0.917. Newly formed sea ice, due to its salt content, is more dense, 0.925 being a representative value. The density decreases as the ice freshens. By the time it has shed most of its salt, sea ice is less dense than fresh-water ice, because ice formed in the sea contains more air bubbles. Ice having no salt but containing air to the extent of eight percent by volume (an approximately maximum value for sea ice) has a density of 0.845.

The density of land ice varies over even wider limits. That formed by freezing of fresh water has a density of 0.917, as stated above. Much of the land ice, however, is formed by compacting of snow. This results in the entrapping of relatively large quantities of air. Névé, in the transitional stage between snow and ice, may have an air content of as much as 50 percent by volume. By the time the ice of a glacier reaches the sea, its density approaches that of fresh-water ice. A sample taken from an iceberg on the Grand Banks had a density of 0.899.

When ice floats, part of it is above water and part is below the surface. The percentage of the mass below the surface can be found by dividing the average density of the ice by the density of the water in which it floats. Thus, if an iceberg of density 0.920 floats in water of density 1.028 (corresponding to a salinity of 35 parts per thousand and a temperature of 30° F), 89.5 percent of its mass will be below the surface. That is, about nine-tenths of the mass will be below the surface, and only about one-tenth will be above the surface. If the ice is a perfectly uniform block, which some tabular icebergs approach, the depth below the surface is about seven times the height above water, under the conditions stated above. However, most of the icebergs of the northern hemisphere are irregular in shape, the depth probably averaging about five times the height. Icebergs have been estimated to be as high as 1,000 feet above water, but the highest measured in the northern hemisphere was 447 feet. The largest tabular icebergs of the antarctic extend about 300 feet above the water.

Drift of Ice—Although surface currents have some effect upon the drift of pack ice, the principal factor is wind. Due to Coriolis force, ice does not drift in the direction of the wind, but about 30° from this direction. In the northern hemisphere, this drift is to the *right* of the direction toward which the wind blows, and in the southern hemisphere it is toward the *left*. Since the surface wind is deflected about twice this amount from the direction of the pressure gradient, the total deflection of the ice is about 90° from the pressure gradient, or along the isobars, with the atmospheric low toward the left and the high toward the right in the northern hemisphere. In the southern hemisphere, these directions are reversed. The *rate* of drift is about one to seven percent of the wind speed, depending upon the roughness of the surface and the concentration of the ice.

Icebergs, which extend a considerable distance below the surface, and have a relatively small "sail area," are influenced more by surface currents than by wind. However, if a strong wind blows for a number of hours in a steady direction, the drift of icebergs will be materially affected. In this case the effect is two-fold. The wind acts directly against the iceberg, and also generates a surface current in about the same direction. Because of inertia, an iceberg may continue to move from the influence of wind for some time after the wind stops or changes direction.

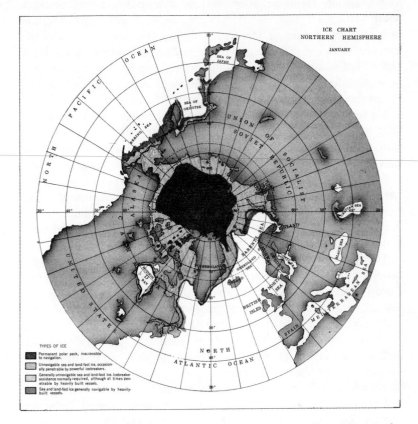

Figure 6–5 *Average limits of various degrees of navigability of ice in the northern hemisphere in January.*

Extent of Ice in the Sea—Several Hydrographic Office publications contain monthly charts showing average extent of various degrees of navigability in the northern and southern hemispheres throughout the year. A sample of the type of information given is shown in figure 6–5 below. Similar information is shown on the various pilot charts. Useful information on ice conditions in different localities is given in the sailing directions for those areas. The information given in H.O. Pub. No. 27, *Sailing Directions, Antarctica*, is particularly complete and of somewhat general application.

However, since formation of ice, in common with other meteorological and oceanographic phenomena, varies considerably from year to year, wide deviations from average conditions are not unusual. Most countries having vessels operating in ice maintain ice information services. Details of these services are given in the appropriate volumes of sailing directions. The ice bulletins broadcast by the U. S. Navy Hydrographic Office are discussed later in this chapter. The latest bulletins, as well as information on average conditions, should be consulted when operating in ice.

Ice in the North Atlantic—Sea-level glaciers exist on a number of land masses bordering the northern seas, including Alaska, Greenland, Svalbard (Spitzbergen), Zemlya Frantsa-Iosifa (Franz Josef Land), Novaya Zemlya, and Severnaya Zemlya (Nicholas II Land). Except in Greenland, the rate of calving is relatively slow, and the few icebergs produced melt near their points of formation. Many of those produced along the coasts of Greenland are eventually carried into the shipping lanes of the North Atlantic, where they constitute a major menace.

The icebergs produced along the east coast of Greenland are carried by the east Greenland current around Kap Farvel and northward by the west Greenland current toward Davis Strait. Relatively few of these icebergs menace shipping, but they have been encountered as far as 200 miles southeast of Kap Farvel.

The most prolific source of icebergs is the west coast of Greenland. In this area there are about 100 tidewater glaciers, 20 of them being the principal producers of icebergs. About 7,500 icebergs are formed here each year. The west Greenland current carries them northward and then westward until they encounter the south-flowing Labrador current. West Greenland icebergs generally spend their first winter in Baffin Bay. During the next summer they are carried southward by the Labrador current. In many cases, their second winter is spent in Davis Strait. When they are freed by the break-up of the pack ice, they drift southward. An average of about 400 per year reach latitude 48° N, and about 35 are carried south of the Grand Banks (latitude 43° N) before they melt. Icebergs have been encountered south of Bermuda, off the Azores, and within a few hundred miles of Great Britain.

The variation from average is considerable. More than 1,350 icebergs have been sighted south of latitude 48° N in a single year (1929), while in 1940 only two were encountered in this area. Although this variation has not been fully explained, it is apparently related to wind conditions, the distribution of pack ice in Davis Strait, and to the amount of pack ice off Labrador. It has been suggested that the distribution of the Davis Strait-Labrador Sea pack ice in-

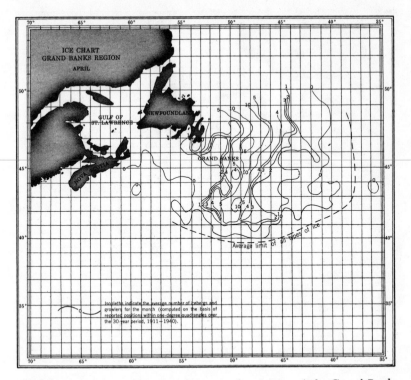

Figure 6–6 *Average iceberg conditions in the vicinity of the Grand Banks in April.*

fluences the effectiveness of this ice in holding back the icebergs. According to this theory, when pack ice is heavy along the Labrador coast, the icebergs are forced well offshore, where warmer water causes them to melt before they reach the North Atlantic shipping lanes; but when the pack ice is not sufficient for this, the icebergs drift closer to shore, where there is colder water which prolongs their existence.

Icebergs may be encountered during any part of the year, but in the Grand Banks area they are most numerous during the spring. Average iceberg and pack ice conditions in this area during April, May, and June are shown in figures 6–6, 6–7, and 6–8. Off Newfoundland, part of the pack ice is brought south by the Labrador current, and part of it comes through Cabot Strait, having originated in the Gulf of St. Lawrence.

The North Atlantic Lane Routes—In his 1855 sailing directions,

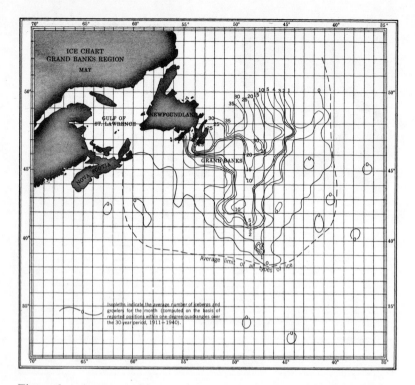

Figure 6–7 *Average iceberg conditions in the vicinity of the Grand Banks in May.*

Matthew Fontaine Maury, who first published the *Wind and Current Chart of the North Atlantic* in 1847, included a section on "Steam Lanes Across the Atlantic." Maury was inspired by the collision and sinking the previous year of the French *Vesta* and American *Arctic*, in which about 300 lives were lost. He recommended separate routes for eastbound and westbound vessels to avoid the risks due to fog. The U. S. Navy Hydrographic Office continued to advocate the use of lane routes during the next 35 years, ultimately designating different routes for different times of the year to avoid ice dangers. In 1889 representatives of 26 maritime nations, meeting at the International Marine Conference in Washington, ruled against establishing steamer lanes by international agreement of the governments involved, but recommended that companies engaged in the North Atlantic trade establish such routes for their own vessels. Two years later

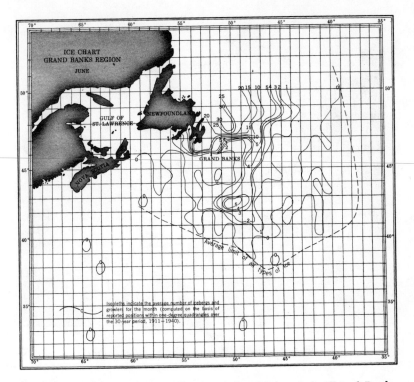

Figure 6–8 *Average iceberg conditions in the vicinity of the Grand Banks in June.*

a group of steamship companies operating passenger liners in the North Atlantic, led by the Cunard Line, agreed to follow designated tracks which were essentially the ones proposed by the U. S. Navy Hydrographic Office. The lanes have been altered somewhat from time to time. The principal ones now in use are shown in figure 6–9. Each lane is composed of two tracks separated by a safe distance, the southern track being used by eastbound vessels, and the northern one by westbound vessels.

Routes *A*, *B*, and *C* connect the United States and Europe, while routes *D*, *E*, *F*, and *G* run between Canada and Europe. Normally, route *B* is used between April 11 and June 30, and route *C* during the remainder of the year. However, when icebergs are numerous south of the Grand Banks, the use of lane *A* is specified. This route adds 150 to 200 miles to the great-circle track, but the increased

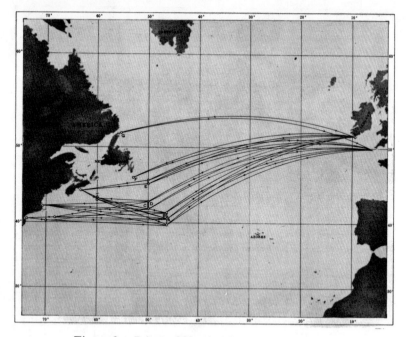

Figure 6–9 *Principal North Atlantic steamer lanes.*

distance is acceptable because it improves the safety and reduces the possibility of delays due to pack ice. Normally, route *D* is used between February 15 and April 10; route *E* from April 11 through May 15, and from December 1 through February 14; route *F* from May 16 until route *G* is clear, about July 1, or through November 30 if route *G* is not used; and route *G* from the opening of Belle Isle Strait, about July 1, through November 14. Specified lanes are shown on pilot charts for the North Atlantic (H.O. Chart No. 1400).

Variations in this schedule are recommended by the Hydrographer of the U. S. Navy. The Hydrographer makes his recommendation after consultation with the Commandant of the U. S. Coast Guard, taking into account the information provided by the International Ice Patrol. Virtually all passenger liners and most freight vessels use these routes.

The International Ice Patrol was established in 1913 by the International Convention for the Safety of Life at Sea held that year as a result of the sinking of the SS *Titanic* the previous year. On its maiden voyage this vessel struck an iceberg and sank with the loss of 1,513 lives. The patrol is conducted by the U. S. Coast Guard, which each year assigns vessels to the vicinity of the Grand Banks during the ice season to observe and report ice conditions.

During the war years of 1916–18 and 1941–45 the patrol was suspended. Following World War II, aircraft were added to the patrol force, and Argentia, Newfoundland, was established as the base of operations. Aircraft have played an increasing role in ice reconnaissance each year since then, and today they perform most of the work. Twice each day during the iceberg season an ice bulletin is broadcast from Argentia and printed in the *Daily Memorandum* of the U. S. Navy Hydrographic Office. Ice patrol vessels copy the broadcasts when on station and make them available to other ships upon request. In return for this service, all vessels in the area are requested to report to the patrol vessels any ice observed, and to send weather data and surface sea water temperature every four hours.

When engaged in patrolling ice areas, the vessels conduct oceanographic surveys and maintain an up-to-date map of the currents, for use in predicting future drift of icebergs. Recommendations for changes in the use of lane routes are based on information gathered by the International Ice Patrol.

Ice Detection—As a ship proceeds into higher latitudes, the first ice it encounters is likely to be in the form of icebergs, because such large pieces require a longer time to disintegrate. Icebergs can easily be avoided if detected soon enough. The distance at which an iceberg can be seen depends upon the visibility, height of the berg, source and condition of lighting, and the observer. On a clear day with excellent visibility a large berg might be sighted at a distance of 18 miles. With a low-lying haze around the horizon this may be reduced to ten miles. In light fog or drizzling rain this is further reduced to one to three miles. There is a tendency to overestimate the distance.

In a dense fog a berg may not be visible until it is close aboard, when it appears as a luminous, white mass if the sun is shining; or as a dark, sombre mass if the sun is not shining. If the layer of fog is not thick, an iceberg may be sighted from aloft sooner than from a point lower in the vessel, but this fact should not be considered justification for omitting a bow lookout.

On a clear, dark night an iceberg will seldom be picked up visually at a distance greater than one-fourth of a mile, but if its bearing is known, an observer with binoculars can occasionally observe a light spot where a wave breaks against it at a distance of a mile.

A moon may either help or hinder, depending upon its phase and position relative to ship and berg. A full moon in the direction of the berg interferes with its detection, while light from one in the opposite direction produces a "blink" which renders the iceberg visible for a greater distance, possibly as much as three miles. Clouds, particularly broken clouds, with intermittent moonlight, add to the difficulty of detecting ice.

If an iceberg is in the process of disintegration, its presence may be detected by the cracking sound as a piece breaks off, or by the thunderous roar as a large piece falls into the water. The appearance of smaller pieces of ice in the water often indicates the presence of an iceberg nearby. In calm weather such pieces may form a curved line with the parent iceberg on the concave side. Some of the pieces broken from an iceberg are themselves large enough to be a menace to ships.

As the ship proceeds to higher latitudes, it eventually encounters pack ice. If the ice is approached from leeward, it is likely to be loose and somewhat scattered, often in long, narrow arms. If it is approached from windward, it is usually compact and the edge is sharply defined.

One of the most reliable signs of the approach to pack ice, especially from leeward, is the somewhat abrupt smoothing of the sea in a fresh breeze, and the more gradual lessening of the swell. Abrupt changes in air or sea temperature or sea-water salinity are not reliable signs of the approach to either icebergs or pack ice, but if the sea temperature gradually drops below 32° F, the ship *may* be nearing an ice field.

Another reliable sign of the approach to pack ice is the appearance of the horizon or sky. A yellowish glare or **ice blink** appears in the sky above an ice field. If clouds are present, the blink is whiter. Reflection of light from snow, whether on land or sea ice, is white and is called **snow blink.** In contrast, the sky above open water is dark. This is called **water sky.** Somewhat similar **land sky** above ice- and snow-free land is grayer. The combination of these various effects in the sky is called a **sky map.** One experienced in reading the sky map finds it very useful in avoiding ice or searching out openings which may permit his vessel to make progress while proceeding through an ice field.

The presence of seals or certain types of birds may indicate the presence of ice nearby. It is well to observe the habits of the various species encountered.

Echoes from a whistle or horn will sometimes reveal the presence of icebergs, but are useless against pack ice. Such echoes can give an indication of direction, and if the time interval between the sound and its echo is measured, the distance in feet can be determined by multiplying the number of seconds by 550. However, echoes are not a reliable indication because only those pieces of ice with large vertical areas facing the ship return enough echo to be heard, and also because echoes might be received from land or a fog bank.

At relatively short ranges, sonar is sometimes helpful in locating ice. The first contact with icebergs may be when as much as three miles or more off, but is usually considerably less. Growlers may be picked up at one-half to one mile and even smaller pieces may be detected in time to avoid them. Since one-half to seven-eights of the mass of ice is below the surface, the underwater portion presents a better target than the portion above water.

Radar is highly useful in detecting ice, but is by no means infallible. Ice is a relatively poor radar target, and much depends upon the nature of the exposed surface. Icebergs with sides sloping gently toward the vessel can be seen visually long before they are picked up by radar, if the day is clear. One iceberg 700 feet long and 200 feet high was reported to have been approached to within three miles before it appeared on the radar screen. However, the average berg is picked up at a range of eight to ten miles, and the large vertical-sided tabular icebergs of the antarctic are usually detected at ranges of 15 to 30 miles, with an extreme range of 37 miles having been reported. Growlers are the chief concern. While a large iceberg is almost always detected in time to be avoided, a growler large enough to be a serious menace may be lost in the sea return and escape detection altogether. If an iceberg or growler is detected, tracking is sometimes necessary to distinguish it from a rock, islet, or ship.

Against sea ice, radar can be of great assistance to one experienced in interpreting the scope picture. Smooth sea ice, like smooth water, returns little or no echo, but rough, hummocky sea ice can detected at a range of two to three miles. The return is similar to sea return, but the same echoes appear at each sweep. A lead in smooth ice broken by a preceding vessel is clearly visible, even though a thin coating of new ice has formed in the opening. A light covering of snow obliterating many of the features to the eye has little effect upon a radar return.

The ranges at which ice can be detected by radar are somewhat dependent upon refraction, which is sometimes quite abnormal in polar regions. Adequate training and experience are essential if full benefit is to be realized from radar.

No method yet devised to detect the presence of ice is infallible, and all should be regarded with suspicion, although none should be overlooked. In ice, as elsewhere, *there is no substitute for constant vigilance.*

In the vicinity of icebergs, a sharp lookout should be kept and all bergs given a wide berth. It is dangerous to approach close to them because of the possibility of encountering underwater extensions and because bergs that are disintegrating may suddenly capsize or readjust their masses to new positions of equilibrium. In periods of low visibility the utmost caution is needed. The speed should be reduced and the watch prepared for quick maneuvering.

Upon the approach to pack ice, a careful decision is needed to determine the best action. Often it is possible to go around the ice, rather than through it. Unless the pack is quite loose, this action usually gains rather than loses time. When skirting a field of ice or an iceberg, do so to windward, if a choice is available, to avoid projecting tongues of ice or individual pieces that have been blown away from the main body of ice.

When it is considered necessary to enter pack ice, select the point of entry with great care. Get all available information on the nature and extent of ice and open water. Seek the weakest part of the ice and particularly avoid ice under pressure. If an offshore wind is blowing, a relatively ice-free shore lead may be available. Enter ice from leeward if possible, at slow speed. Enter on a course perpendicular to the ice edge, avoiding projecting tongues of ice.

Having entered the pack, always *work with the ice, not against it,* and *keep moving*, but do not rush the work of negotiating the pack. Patience may pay big dividends. Respect the ice but do not fear it. Stay in open water or areas of weak ice if possible, remembering that it is better to make good progress in the *general* direction desired than to fight heavy floes in the *exact* direction to be made good. However, avoid the temptation to proceed far to one side of the course. It is sometimes better to back out and seek a more penetrable area, being careful not to damage the screws while backing. Keep clear of corners and projecting points of ice. Never hit a large piece of ice if it can be avoided, but if it cannot be avoided, hit it head-on. Keep a sharp watch on the screws and rudder, fending off pieces of ice which might damage these vital parts, or stopping the propellers if the ice cannot be avoided. Back with extreme caution.

The windward side of icebergs within pack ice should be avoided because the pack ice usually moves with the wind, while the berg does not do so to the same extent, resulting in pressure on the windward side and open water to leeward. Because of its poor maneuver-ability in ice, a vessel may even be set down upon the iceberg.

If a narrow strait or a bay is entered, an alert watch should be maintained, because if the wind blows directly into the confined space, drifting ice may be forced down upon the vessel. An increase in wind on the windward side of a prominent point, grounded iceberg, or land ice tongue extending into the sea may similarly endanger a vessel.

While a vessel is in pack ice, it is always in danger of being **beset,** or so closely surrounded by ice that steering control is lost. It may then be carried into shallow water or heavy ice with dangerous under-water projections. If pressure is exerted against the hull, the vessel is said to be **nipped.** When this occurs, it is in danger of being crushed. A ship in the ice is in constant danger of colliding with sharp pieces of ice, and while in the ice sharp turns to avoid such collisions may throw the stern against the ice, resulting in a bent or broken screw blade or propeller shaft.

Only the basic principles of operating in ice have been given. Before entering areas of ice, those responsible for the maneuvering of a ship should become well acquainted with the experience of others who have operated in ice, especially those who have been in the same area. Some of this information is to be found in various volumes of sailing directions, particularly those for Antarctica (H.O. Pub. No. 27), and additional information is available at the U.S. Navy Hydro-graphic Office.

Ice Observing and Forecasting—Advance knowledge of ice con-ditions to be encountered is valuable in both planning and operational phases of any program to be conducted in high latitudes. Through the cooperation of observers aboard ship, in the air, and on land, the U. S. Navy Hydrographic Office collects and analyzes ice data in the arctic, and distributes ice information in the form of ice bulletins as part of regularly scheduled broadcasts.

For this program to be fully effective, it is essential that all vessels operating in ice areas cooperate by submitting reports. To assist in this program, and to provide uniformity in reporting procedure, the U. S. Navy Hydrographic Office has published an observer's manual, H.O. Pub. No. 606–d, *Ice Observations;* H.O. Pub. No. 609, *A Functional Glossary of Ice Terminology;* and convenient ice log forms for recording the observations. When filled in, the log sheets are mailed to the U. S. Navy Hydrographic Office, Washington, D. C., and certain reports are sent by radio. The mariner who regularly sends complete reports can contribute to an increase in knowledge of ice conditions and to the accuracy and completeness of ice bulletins.

The Sky

=7=

Weather and Weather Forecasts

Weather is the state of the earth's atmosphere with respect to temperature, humidity, precipitation, visibility, cloudiness, etc. In contrast, the term **climate** refers to the prevalent or characteristic meteorological conditions of a place or region.

All weather may be traced ultimately to the effect of the sun on the earth, including the lower portions of the atmosphere. Most changes in weather involve large-scale, approximately horizontal, motion of air. Air in such motion is called **wind.** This motion is produced by differences of atmospheric pressure, which are largely attributable to differences of temperature.

The weather is of vital interest to the mariner. The wind and state of the sea affect dead reckoning. Reduced horizontal visibility limits piloting. The state of the atmosphere affects electronic navigation and radio communication. If the skies are overcast, visual celestial observations are not available; and under certain conditions refraction and dip are disturbed. When wind was the primary motive power, knowledge of the areas of favorable winds was of great importance. This consideration led Matthew Fontaine Maury, more than a century ago, to seek information from ships' logs to establish speed and direction of prevailing winds over the various trade routes of the world. The information thus gathered was shown on pilot charts. By means of these charts, the mariner could select a suitable route for a favorable passage. Even power vessels are affected considerably by wind and sea. Less fuel consumption and a more comfortable passage are to be expected if wind and sea are moderate and favorable. Pilot charts are useful in selecting suitable routes. Since longer range forecasts have become possible, some experimental work has been done in routing ocean vessels to take advantage of anticipated conditions during passage.

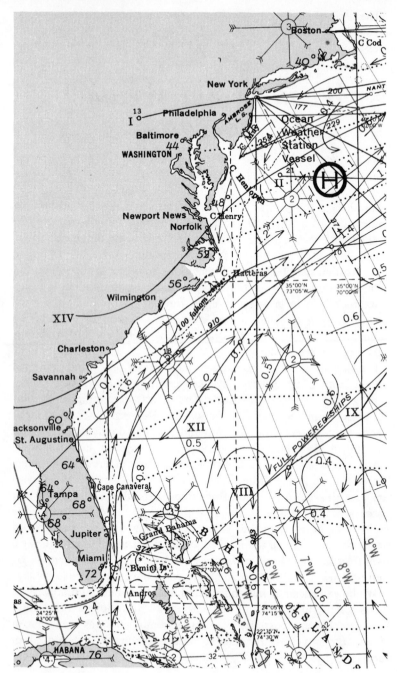

The atmosphere is a relatively thin shell of air, water vapor, dust, and smoke surrounding the earth. The air is a mixture of transparent gases and, like any gas, is elastic and highly compressible. Although extremely light, it has a weight which can be measured. A cubic foot of air at standard sea-level temperature and pressure weighs 1.22 ounces, or about 1/817th part of the weight of an equal volume of water. Because of this weight, the atmosphere exerts a pressure upon the surface of the earth, amounting to about 15 pounds per square inch.

As altitude increases, less atmosphere extends upward, and pressure decreases. With less pressure, the density decreases. More than three-fourths of the air is concentrated within a layer averaging about seven statute miles thick, called the **troposphere.** This is the region of most "weather," as the term is commonly understood.

The top of the troposphere is marked by a thin transition zone called the **tropopause.** Beyond this lie several other layers having distinctive characteristics. The average height of the tropopause ranges from about five miles or less over the poles to about 11 miles over the equator.

The **standard atmosphere** is a conventional vertical structure of the atmosphere characterized by standard sea level pressure of 29.92 inches of mercury (1013.25 millibars), sea level temperature of 59° F (15° C), and a uniform decrease of temperature and moisture content of the air with height, the rate of temperature decrease being 3°.6F (2° C) per thousand feet to 11 kilometers (36,089 feet) and thereafter a constant temperature of (−)69°.7F (−56°.5 C). The rate of temperature decrease with height in the standard atmosphere is called the **standard temperature lapse rate.**

Meteorologists are continually learning more of the characteristics of atmospheric processes above the lowest portions of the atmosphere. In recent years, greatly increased attention has been directed to such features as the **jet stream,** a meandering stream of air which circles the globe at speeds of 100 to more than 250 knots at heights of about 20,000 to 40,000 feet. Some similarity has been noted between major wind streams such as the jet stream, and ocean currents such as the Gulf Stream.

Wind—When air is not confined, changes in temperature produce changes in volume, heated air expanding and cooled air contracting. If a large volume of air near the surface of the earth is cooled, it contracts, causing a downdraft. Air from neighboring regions aloft moves horizontally to fill the void. This results in a greater mass of air over the region, and the pressure is correspondingly increased. By a similar process in reverse, heating of air near the surface causes expansion and an updraft, resulting in decreased pressure over the heated area. Near the surface of the earth, the air tends to move from an area of high pressure to one of low pressure. Thus, a circulation is set up, air moving across the surface of the earth from an area of high pressure and low temperature to one of low pressure and high temperature, then vertically upward, then horizontally at high altitudes from the area of low pressure to that of high pressure, where it moves vertically downward to complete the circuit. The actual circulation is much more complex than this, due to such factors as rotation of the earth and continual changes in temperature and pressure.

If there were no heating and cooling, the temperature at any given altitude remaining everywhere the same, there would be no tendency for the air to move from one place to another. Air would lie sluggish and at rest on the earth's surface. There would be no wind and no variation in weather.

As a result of the position and motion of the earth in relation to the sun, and the physical processes involving radiation and absorption of energy, certain regions of the earth are always warmer than others. For similar reasons, the air over some parts of the earth is seasonally warmer than that over other parts. This general pattern is modified to a varying degree by the local heating and cooling which is continually taking place. Consequently, winds in some areas are relatively steady in both direction and speed, others are seasonal, and this general circulation is continually being modified by local conditions.

General Circulation of the Atmosphere—The heat required for warming the air is supplied originally by the sun. As radiant energy from the sun arrives at the earth, about 43 percent is reflected back into space by the atmosphere, about 17 percent is absorbed in the lower portions of the atmosphere, and the remaining 40 percent (approximately) reaches the surface of the earth and much of it is re-radiated into space. This earth radiation is in comparatively long waves relative to the short-wave radiation from the sun, since it emanates from a cooler body. Long-wave radiation, being readily absorbed by the water vapor in the air, is primarily responsible for the warmth of the atmosphere near the earth's surface. Thus, the atmosphere acts much like the glass on the roof of a greenhouse. It allows part of the incoming solar radiation to reach the surface of the earth, but is heated by the terrestrial radiation passing outward. Over the entire earth and for long periods of time, the total outgoing energy must be equivalent to the incoming energy (minus any converted to another form and retained), or the temperature of the earth, including its atmosphere, would steadily increase or decrease. In local areas, or over relatively short periods of time, such a balance is not required, and in fact does not exist, resulting in changes such as those occurring in the different seasons, and in different parts of the day.

The more nearly perpendicular the rays of the sun as they strike the surface of the earth, the more heat energy per unit area is received at that place. Physical measurements show that in the tropics more heat per unit area is received than is radiated away, and that in polar regions the opposite is true. Unless there were some process to transfer heat from the tropics to polar regions, the tropics would be much warmer than they are, and the polar regions would be much colder. The process which brings about the required transfer of heat is the general circulation of the atmosphere.

If the earth had a uniform surface, did not rotate on its axis (but received sunlight equally all around the equator), and did not revolve around the sun (with its axis tilted), a simple circulation would result, as shown in figure 7–1. However, the surface of the earth is far from uniform, being covered with an irregular distribution of land of various heights, and water; the earth rotates about its axis once in approximately 24 hours, so that the portion heated by the sun continually changes; and the axis of rotation is tilted so that as the earth moves along its orbit about the sun, seasonal changes occur in the exposure of specific areas to the sun's rays, resulting in variations in the heat balance of these areas. These factors, coupled with others, result in constantly changing large-scale movements of air. Based upon averages over long periods, however, a general circulation is discernible. Figures 7–2 and 7–3 give a generalized picture of the world's pressure distribution and wind systems as actually observed. A simplified diagram of the general pattern is shown in figure 7–4.

The rotation of the earth diverts the air from a direct path between high and low pressure areas, the diversion being toward the *right* in the northern hemisphere and toward the *left* in the southern hemisphere. At some distance above the surface of the earth, the wind tends to blow along the isobars, being called the **geostrophic wind** if the isobars are straight (great circles), and **gradient wind** if they are curved. Near the surface of the earth, friction tends to divert the wind from the isobars toward the center of low pressure. At sea, where friction is less than on land, the wind follows the isobars more closely.

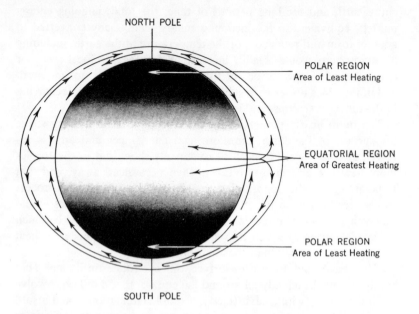

NORTH POLE

POLAR REGION
Area of Least Heating

EQUATORIAL REGION
Area of Greatest Heating

POLAR REGION
Area of Least Heating

SOUTH POLE

Figure 7–1 *Ideal atmospheric circulation for a uniform, nonrotating, non-revolving earth.*

The decrease of pressure with distance is called the **pressure gradient.** It is maximum along a normal (perpendicular) to the isobars, decreasing to zero along the isobars. Speed of the wind is directly proportional to the maximum pressure gradient.

The Doldrums—The belt of low pressure near the equator occupies a position approximately midway between high pressure belts at about latitude 30° to 35° on each side. Except for slight diurnal changes, the atmospheric pressure along the equatorial low is almost uniform. With almost no pressure gradient, wind is practically non-existent. The light breezes that do blow are variable in direction. Hot, sultry days are common. The sky is often overcast, and showers and thundershowers are relatively frequent.

The area involved is a thin belt near the equator, the eastern part in both the Atlantic and Pacific being wider than the western part. However, both the position and extent of the belt vary somewhat with the season. During February and March it lies immediately to the north of the equator and is so narrow that it may be considered virtually nonexistent. In July and August the belt is centered on about

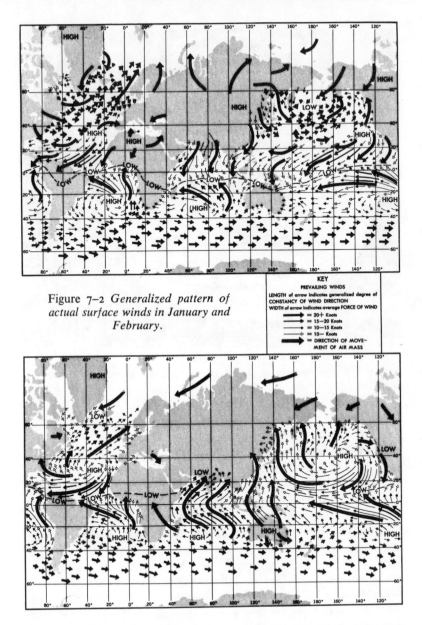

Figure 7–2 *Generalized pattern of actual surface winds in January and February.*

KEY
PREVAILING WINDS
LENGTH of arrow indicates generalized degree of CONSTANCY OF WIND DIRECTION
WIDTH of arrow indicates average FORCE OF WIND
= 20+ Knots
= 15—20 Knots
= 10—15 Knots
= 10— Knots
= DIRECTION OF MOVEMENT OF AIR MASS

Figure 7–3 *Generalized pattern of actual surface winds in July and August.*

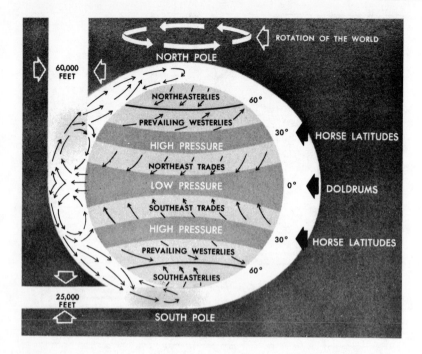

Figure 7–4 *Simplified diagram of the general circulation of the atmosphere.*

latitude 7° N, and is several degrees in width, even at the narrowest point.

The Trade Winds—The trades blow from the belts of high pressure, toward the equatorial belt of low pressure. Because of the rotation of the earth, the moving air is deflected toward the west. Therefore, the trade winds in the northern hemisphere are from the northeast and are called the **northeast trades,** while those in the southern hemisphere are from the southeast and are called the **southeast trades.** Over the eastern part of both the Atlantic and Pacific these winds extend considerably farther from the equator, and their original direction is more nearly along the meridians than in the western part of each ocean.

The trade winds are generally considered among the most constant of winds. Although they sometimes blow for days or even weeks with little change of direction or speed, their constancy is sometimes exaggerated. At times they weaken or shift direction, and there are regions where the general pattern is disrupted. A notable example is the island groups of the South Pacific, where they are practically nonexistent during January and February. Their highest development is attained in the South Atlantic and in the South Indian Ocean. Everywhere they are fresher during the winter than during the summer.

In July and August, when the belt of equatorial low pressure moves to a position some distance north of the equator, the southeast trades blow across the equator, into the northern hemisphere, where the earth's rotation diverts them toward the right, causing them to be southerly and southwesterly winds. The "southwest monsoons" of the African and Central American coasts have their origin partly in such diverted southeast trades.

Cyclonic storms generally do not enter the regions of the trade winds, although hurricanes and typhoons may originate within these areas.

The Horse Latitudes—Along the poleward side of each trade-wind belt, and corresponding approximately with the belt of high pressure in each hemisphere, is another region with weak pressure gradients and correspondingly light, variable winds. These are called the **horse latitudes.** The weather is generally clear and fresh, unlike that in the doldrums, and periods of stagnation are less persistent, being of a more intermittent nature. The difference is due primarily to the fact that rising currents of warm air in the equatorial low carry large amounts of moisture which condenses as the air cools at higher levels, while in the horse latitudes the air is apparently descending and becoming less humid as it is warmed at lower heights.

The Prevailing Westerlies—On the poleward side of the high pressure belt in each hemisphere the atmospheric pressure again diminishes. The currents of air set in motion along these gradients toward the poles are diverted by the earth's rotation toward the east, becoming southwesterly winds in the northern hemisphere and northwesterly in the southern hemisphere. These two wind systems are known as the **prevailng westerlies** of the temperate zones.

In the northern hemisphere this relatively simple pattern is distorted considerably by secondary wind circulations, due primarily to the presence of large land masses. In the North Atlantic, between latitudes 40° and 50°, winds blow from some direction between south and northwest during 74 percent of the time, being somewhat more persistent in winter than in summer. They are stronger in winter, too, averaging about 25 knots (Beaufort 6) as compared with 14 knots (Beaufort 4) in the summer.

In the southern hemisphere the westerlies blow throughout the year with a steadiness approaching that of the trade winds. The speed, though variable, is generally between 17 and 27 knots (Beaufort 5 and 6). Latitudes 40° S to 50° S (or 55° S), where these boisterous winds occur, are called the **roaring forties.** These winds are strongest at about latitude 50° S.

The greater speed and persistence of the westerlies in the southern hemisphere are due to the difference in the atmospheric pressure pattern, and its variations, from that of the northern hemisphere. In the comparatively landless southern hemisphere, the average yearly atmospheric pressure diminishes much more rapidly on the poleward side of the high pressure belt, and has fewer irregularities due to continental interference, than in the northern hemisphere.

Winds of Polar Regions—Because of the low temperatures near the geographical poles of the earth, the pressure tends to remain higher than in surrounding regions. Consequently, the winds blow outward from the poles, and are deflected westward by the rotation of the earth, to become **northeasterlies** in the arctic, and **southeasterlies** in the antarctic. Where these meet the prevailing westerlies, the winds are variable.

In the arctic, the general circulation is greatly modified by surrounding land masses. Winds over the Arctic Ocean are somewhat variable, and strong surface winds are rarely encountered.

In the antarctic, on the other hand, a high central land mass is surrounded by water, a condition which augments, rather than diminishes, the general circulation. The high pressure, although weaker than in some areas, is stronger than in the arctic, and of great persistence near the south pole. The upper air descends over the high continent, where it becomes intensely cold. As it moves outward and downward toward the sea, it is deflected toward the west by the earth's rotation. The winds remain strong throughout the year, frequently attaining hurricane force, and sometimes reaching speeds of 100 to 200 knots at the surface. These are the strongest surface winds encountered anywhere in the world, with the possible exception of those in well-developed tropical cyclones.

Modifications of the General Circulation—The general circulation of the atmosphere is greatly modified by various conditions. The high pressure in the horse latitudes is not uniformly distributed around the belts, but tends to be accentuated at several points. These **semipermanent highs** remain at about the same places with great persistence.

Semipermanent lows also occur in various places, the most prominent ones being west of Iceland, and over the Aleutians (winter only) in the northern hemisphere, and at the Ross Sea and Weddell Sea in the antarctic. The areas occupied by these semipermanent lows are

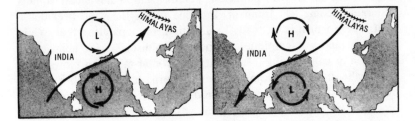

Figure 7–5 *The summer monsoon.*   Figure 7–6 *The winter monsoon.*

sometimes called the graveyards of the lows, since many lows move directly into these areas and lose their identity as they merge with and reinforce the semipermanent lows. The low pressure in these areas is maintained largely by the migratory lows which stall there, but partly by the sharp temperature difference between polar regions and warmer ocean areas.

Another modifying influence is land, which undergoes greater temperature changes than does the sea. During the summer, a continent is warmer than its adjacent oceans. Therefore, low pressures tend to prevail over the land. If a belt of high pressure encounters such a continent, its pattern is distorted or interrupted. A belt of low pressure is intensified. The winds associated with belts of high and low pressure are distorted accordingly. In winter, the opposite effect takes place, belts of high pressure being intensified over land and those of low pressure being interrupted.

The most striking example of a wind system produced by the alternate heating and cooling of a land mass is the **monsoons** of the China Sea and Indian Ocean. A portion of this effect is shown in figures 7–5 and 7–6. In the summer, low pressure prevails over the warm continent of Asia, and high pressure over the adjacent sea. Between these two systems the wind blows in a nearly steady direction. The lower portion of the pattern is in the southern hemisphere, extending to about 10° south latitude. Here the rotation of the earth causes a deflection to the left, resulting in southeasterly winds. As they cross the equator, the deflection is in the opposite direction, causing them to curve toward the right, becoming southwesterly winds. In the winter, the positions of high and low pressure areas are interchanged, and the direction of flow is reversed.

In the China Sea the summer monsoon blows from the southwest, usually from May to September. The strong winds are accompanied by heavy squalls and thunderstorms, the rainfall being much heavier than during the winter monsoon. As the season advances, squalls and rain become less frequent. In some places the wind becomes a light breeze which is unsteady in direction, or stops altogether, while in other places it continues almost undiminished, with changes in direction or calms being infrequent. The winter monsoon blows from the northeast, usually from October to April. It blows with a steadiness similar to that of the trade winds, often attaining the speed of a moderate gale (28–33 knots). Skies are generally clear during this season, and there is relatively little rain.

The general circulation is further modified by winds of cyclonic origin and various local winds.

Air Masses—Because of large differences in physical characteristics of the earth's surface, particularly the oceanic and continental contrasts, the air overlying these surfaces acquires differing values of temperature, moisture, etc. The processes of radiation and convection in the lower portions of the troposphere act in differing, characteristic manners for a number of well-defined regions of the earth. The air overlying these regions acquires characteristics common to the particular area, but contrasting to those of other areas. Each distinctive part of the atmosphere, within which common characteristics prevail over a reasonably large area, is called an **air mass.**

Air masses are named according to their source regions. Four such regions are generally recognized: (1) *equatorial* (*E*), the doldrum area between the north and south trades; (2) *tropical* (*T*), the trade wind and lower temperate regions; (3) *polar* (*P*), the higher temperate latitudes; and (4) *arctic* (*A*), the north polar region of ice and snow (or, by extension, the antarctic). This classification is a general indication of relative temperature, as well as latitude of origin.

Tropical and polar air masses are further classified as *maritime* (*m*) or *continental* (*c*), depending upon whether they form over water or land. This classification is an indication of the relative moisture content of the air mass. Since the moisture content of equatorial and arctic air is essentially independent of the surface over which they form, these sub-classifications are not applied to them. Tropical air, then, might be designated *maritime tropical* (*mT*) or *continental tropical* (*cT*). Similarly, polar air may be either *maritime polar* (*mP*) or *continental polar* (*cP*).

A third classification sometimes applied to tropical and polar air masses indicates whether the air mass is *warm* (*w*) or *cold* (*k*) relative to the underlying surface. Thus, the symbol *mTw* indicates maritime tropical air which is warmer than the underlying surface, and

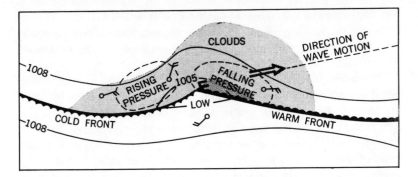

Figure 7–7 *First stage in the development of a frontal wave (top view).*

cPk indicates continental polar air which is colder than the underlying surface. The *w* and *k* classifications are primarily indications of stability. If the air is cold relative to the surface, the lower portion of the air mass is being heated, resulting in instability as the warmer air tends to rise by convection. Conversely, if the air is warm relative to the surface, the lower portion of the air mass is cooled, tending to remain close to the surface. This is a stable condition.

Two other types of air masses are sometimes recognized. These are *monsoon* (*M*), a transitional form between *cP* and *E*; and *superior* (*S*), a special type formed in the free atmosphere by the sinking and consequent warming of air aloft.

Fronts—As air masses move within the general circulation, they travel from their source regions and invade other areas dominated by air having different characteristics. There is little tendency for adjacent air masses to mix. Instead, they are separated by a thin zone in which air mass characteristics exhibit such sharp gradients as to appear as discontinuities. This is called a **frontal surface.** The intersection of a frontal surface and a horizontal plane is called a **front,** although the term "front" is commonly used as a short expression for "frontal surface" when this will not introduce an ambiguity.

Because of differences in the motion of adjacent air masses, "waves" form along the frontal surface between them.

Before the formation of frontal waves, the isobars (lines of equal atmospheric pressure) tend to run parallel to the fronts. As a wave is formed, the pattern is distorted somewhat, as shown in figure 7–7. In this illustration, colder air is north of warmer air. Isobars are shown at intervals of three millibars. The wave tends to travel in the direction of the general circulation, which in the temperate latitudes is usually in a general easterly and slightly poleward direction.

Along the leading edge of the wave, warmer air is replacing colder air. This is called the **warm front.** The trailing edge is the **cold front,** where colder air is replacing warmer air.

The warm air, being less dense, tends to ride up over the colder air it is replacing, causing the warm front to be tilted in the direction of motion. The slope is gentle, varying between 1:100 and 1:300. Because of the replacement of cold, dense air with warm, light air, the pressure decreases. Since the slope is gentle, the upper part of a warm frontal surface may be many hundreds of miles ahead of the surface portion. The decreasing pressure, indicated by a "falling barometer," is often an indication of the approach of such a wave. In a slow-moving, well-developed wave, the barometer may begin to fall several *days* before the wave arrives. Thus, the amount and nature of the change of atmospheric pressure between observations, called **pressure tendency,** is of assistance in predicting the approach of such a system.

The advancing cold air, being more dense, tends to cut under the warmer air at the cold front, lifting it to greater heights. The slope here is in the opposite direction, at a rate of about 1:25 to 1:100, being steeper than the warm front. Therefore, after a cold front has passed, the pressure increases—a "rising barometer."

In the first stages, these effects are not marked, but as the wave continues to grow, they become more pronounced, as shown in figure 7–8. As the amplitude of the wave increases, pressure near the center usually decreases, and the "low" is said to "deepen." As it deepens, its forward speed generally decreases.

The approach of a well-developed warm front is usually heralded not only by falling pressure, but also by a more-or-less regular sequence of clouds. First, cirrus appear. These give way successively to cirrostratus, altostratus, altocumulus, and nimbostratus. Brief showers may precede the steady rain accompanying the nimbostratus.

As the warm front passes, the temperature rises, the wind shifts to the right (in the northern hemisphere), and the steady rain stops. Drizzle may fall from low-lying stratus clouds, or there may be fog for some time after the wind shift. During passage of the **warm sector** between the warm front and the cold front, there is little change in temperature or pressure. However, if the wave is still growing and the low deepening, the pressure might slowly decrease. In the warm sector the skies are generally clear or partly cloudy, with cumulus or

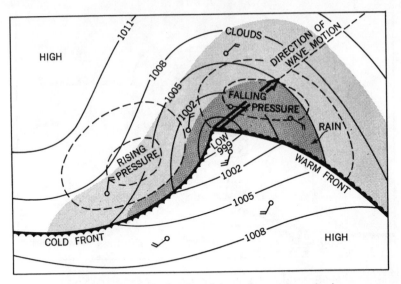

Figure 7–8 *A fully developed frontal wave (top view).*

stratocumulus clouds most frequent. The warm air is usually moist, and haze or fog may often be present.

As the faster moving, steeper cold front passes, the wind shifts abruptly to the right (in the northern hemisphere), the temperature falls rapidly, and there are often brief and sometimes violent showers, frequently accompanied by thunder and lightning. Clouds are usually of the convective type. A cold front usually coincides with a well-defined **wind-shift line** (a line along which the wind shifts abruptly from southerly or southwesterly to northerly or northwesterly in the northern hemisphere and from northerly or northwesterly to southerly or southwesterly in the southern hemisphere). At sea a series of brief showers accompanied by strong, shifting winds may occur along or some distance (up to 200 miles) ahead of a cold front. These are called **squalls,** and the line along which they occur is called a **squall line.** Because of its greater speed and steeper slope, which may approach or even exceed the vertical near the earth's surface (due to friction), a cold front and its associated weather pass more quickly than a warm front. After a cold front passes, the pressure rises, often quite rapidly, the visibility usually improves, and the clouds tend to diminish.

As the wave progresses and the cold front approaches the slower

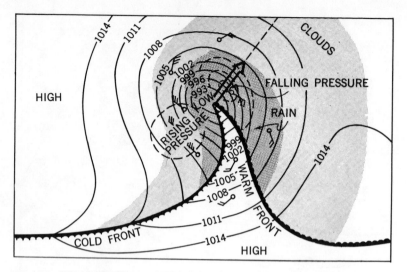

Figure 7–9 *A frontal wave nearing occlusion (top view).*

moving warm front, the low becomes deeper and the warm sector becomes smaller. This is shown in figure 7–9.

Finally, the faster moving cold front overtakes the warm front
(figure 7–10), resulting in an **occluded front** at the surface, and an
upper front aloft, (figure 7–11). When the two parts of the cold air
mass meet, the warmer portion tends to rise above the colder part.
The warm air continues to rise until the entire system dissipates. As
the warmer air is replaced by colder air, the pressure gradually rises,
a process called "filling." This usually occurs within a few days after
an occluded front forms, but the process is sometimes delayed by a
slowing of the forward motion of the wave. In general, however, a
filling low increases in speed.

The sequence of weather associated with a low depends greatly
upon location with respect to the path of the center. That described
above assumes that the observer is so located that he encounters each
part of the system. If he is poleward of the path of the center of the
low, the abrupt weather changes associated with the passage of fronts
are not experienced. Instead, the change from the weather character-
istically found ahead of a warm front to that behind a cold front
takes place gradually, the exact sequence being dictated somewhat by
distance from the center, as well as severity and age of the low.

Although each low follows generally the pattern given above, no

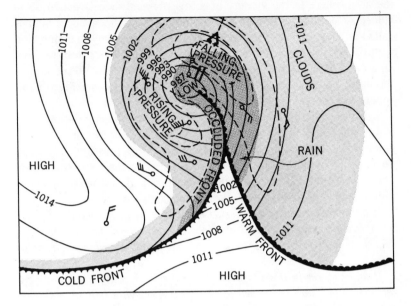

Figure 7–10 *An occluded front (top view).*

two are ever exactly alike. Other centers of low pressure and high pressure and the air masses associated with them, even though they may be 1,000 miles or more away, influence the formation and motion of individual low centers and their accompanying weather. Particularly, a high stalls or diverts a low. This is true of temporary highs as well as semipermanent highs.

Cyclones and Anticyclones—An approximately circular portion of

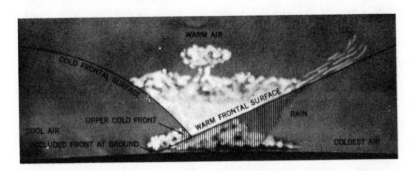

Figure 7–11 *An occluded front (cross section).*

the atmosphere in the vicinity of a low pressure area is called a **cyclone.** A similar portion in the vicinity of an atmospheric high is called an **anticyclone.** These terms are used particularly in connection with the winds associated with such centers. Wind tends to blow from an area of high pressure to one of low pressure, but due to rotation of the earth, they are deflected toward the right in the northern hemisphere and toward the left in the southern hemisphere.

Because of the rotation of the earth, therefore, the circulation tends to be counterclockwise around areas of low pressure in the northern hemisphere, and clockwise around areas of high pressure, the speed being proportional to the spacing of isobars. In the southern hemisphere, the direction of circulation is reversed. Based upon this condition, a general rule **(Buys Ballot's Law)** can be stated thus:

If an observer in the northern hemisphere faces the wind, the center of low pressure is toward his right, somewhat behind him; and the center of high pressure is toward his left and somewhat in front of him.

If an observer in the southern hemisphere faces the wind, the center of low pressure is toward his left and somewhat behind him; and the center of high pressure is toward his right and somewhat in front of him.

In a general way, these relationships apply in the case of the general distribution of pressure, as well as to temporary local pressure systems.

The reason for the wind shift along a front is that the isobars have an abrupt change of direction along these lines. Since the direction of the wind is directly related to the direction of isobars, any change in the latter results in a shift in the wind direction.

In the northern hemisphere, the wind shifts toward the *right* when either a warm or cold front passes. In the southern hemisphere, the shift is toward the *left*. When the wind shifts in this direction (clockwise in the northern hemisphere and counterclockwise in the southern hemisphere), it is said to **veer.** If it shifts in the opposite direction, as when an observer is on the poleward side of the path of a frontal wave, it is said to **back.**

In an anticyclone, successive isobars are relatively far apart, resulting in light winds. In a cyclone, the isobars are more closely spaced. With a steeper pressure gradient, the winds are stronger.

Since an anticyclonic area is a region of outflowing winds, air is drawn into it from aloft. Descending air is warmed, and as air becomes warmer, its capacity for holding uncondensed moisture in-

creases. Therefore, clouds tend to dissipate. Clear skies are characteristic of an anticyclone, although scattered clouds and showers are
sometimes encountered.

In contrast, a cyclonic area is one of converging winds. The resulting upward movement of air results in cooling, a condition favorable to the formation of clouds and precipitation. More or less continuous rain and generally stormy weather are usually associated with
a cyclone.

Between the two belts of high pressure associated with the horse
latitudes, cyclones form only occasionally, generally in certain seasons, and always in certain areas at sea. These **tropical cyclones** are
usually quite violent, being known under various names according to
their location. Tropical cyclones are discussed in Chapter 10.

In the areas of the prevailing westerlies, cyclones are a common
occurrence, the cyclonic and anticyclonic circulation being a prominent feature of temperate latitudes. These are sometimes called
extratropical cyclones to distinguish them from the more violent
tropical cyclones. Although most of them are formed at sea, their
formation over land is not unusual. As a general rule, they decrease
in intensity when they encounter land, and increase when they move
from land to a water area. In their early stages, cyclones are elongated,
but as their life cycle proceeds, they become more nearly circular.

Local Winds—In addition to the winds of the general circulation
and those associated with cyclones and anticyclones, there are numerous local winds which influence the weather in various places.

The most common of these are the **land** and **sea breezes,** caused by
alternate heating and cooling of land adjacent to water. The effect is
similar to that which causes the monsoons, but on a much smaller
scale, and over shorter periods. By day the land is warmer than the
water, and by night it is cooler. This effect occurs along many coasts
during the summer. Between about 0900 and 1100 the temperature
of the land becomes greater than that of the adjacent water. The
lower levels of air over the land are warmed, and the air rises, drawing in cooler air from the sea. This is the **sea breeze.** Late in the afternoon, when the sun is low in the sky, the temperature of the two
surfaces equalizes and the breeze stops. After sunset, as the land
cools below the sea temperature, the air above it is also cooled. The
contracting cool air becomes more dense, increasing the pressure.
This results in an outflow of winds to the sea. This is the **land breeze,**
which blows during the night and dies away near sunrise. Since the
atmospheric pressure changes associated with this cycle are not great,

the accompanying winds do not exceed gentle breezes. The circulation is generally of limited extent, reaching a distance of perhaps 20 miles inland, and not more than five or six miles offshore, and to a height of a few hundred feet. In the tropics, this process is repeated with great regularity throughout most of the year. As the latitude increases, it becomes less prominent, being masked by winds of cyclonic origin. However, the effect may often be present to reinforce, retard, or deflect stronger prevailing winds.

Varying conditions of topography produce a large variety of local winds throughout the world. In light airs, winds tend to follow valleys, and to be deflected from high banks and shores. Many local winds have been given distinctive names. An **anabatic wind** is one which blows up an incline, as one which blows up a hillside due to surface heating. A **katabatic wind** is one which blows down an incline due to cooling of the air. The cooler air becomes heavier than surrounding air and flows downward along the incline under the force of gravity.

A dry wind with a downward component, warm for the season, is called a **foehn.** The foehn occurs when horizontally moving air encounters a mountain barrier. As it blows upward to clear the barrier, it is cooled below the dew point, resulting in loss of moisture by cloud formation and perhaps rain. As the air continues to rise, its rate of cooling is reduced because the condensing water vapor gives off heat to the surrounding atmosphere. After crossing the mountain barrier, the air flows downward along the leeward slope, being warmed by compression as it descends to lower levels. Thus, since it loses less heat on the ascent than it gains during descent, and since it loses moisture during ascent, it arrives at the bottom of the mountains as very warm, dry air. This accounts for the warm, arid regions along the eastern side of the Rocky Mountains and in similar areas. In the Rocky Mountain region this wind is known by the name **chinook.** It may occur at any season of the year, at any hour of the day or night, and have any speed from a gentle breeze to a gale. It may last for several days, or for a very short period. Its effect is most marked in winter, when it may cause the temperature to rise as much as 20°F to 30°F within 15 minutes, and cause snow and ice to melt within a few hours. On the west coast of the United States, the name "chinook" is given to a moist southwesterly wind from the Pacific Ocean, warm in winter and cool in summer. Cloudy weather and rain may accompany or follow this wind, which is thus quite different from the other chinook mentioned above. A foehn given the name **Santa Ana** blows through a pass and down a valley by that name in

Southern California. This wind usually starts suddenly, without warn- ing, and blows with such force that it may capsize small craft off the coast.

A cold wind blowing down an incline is called a **fall wind.** Al- though it is warmed somewhat during descent, as is the foehn, it is cold relative to the surrounding air. It occurs when cold air is dammed up in great quantity on the windward side of a mountain and then spills over suddenly, usually as an overwhelming surge down the other side. It is usually quite violent, sometimes reaching hurri- cane force. A different name for this type wind is given at each place where it is common. The **williwaw** of the Aleutian coast, the **tehuante- pecer** of the Mexican and Central American coast, the **pampero** of the Argentine coast, the **mistral** of the western Mediterranean, and the **bora** of the eastern Mediterranean are examples of this type wind.

Many other local winds common to certain areas have been given distinctive names.

A **blizzard** is a violent, intensely cold wind laden with snow mostly or entirely picked up from the ground, although the term is often used popularly to refer to any heavy snowfall accompanied by strong wind. A **dust whirl** is a rotating column of air about 100 to 300 feet in height, carrying dust, leaves, and other light material. This wind, which is similar to a waterspout at sea, is given various local names such as **dust devil** in southwestern United States and **desert devil** in South Africa. A **gust** is a sudden, brief increase in wind speed followed by a slackening, or the violent wind or squall that accompanies a thunderstorm. A puff of wind or a light breeze affecting a small area, such as would cause patches of ripples on the surface of water, is called a **cat's paw.**

Fog, like a cloud, is a visible assemblage of numerous tiny drop- lets of water, or ice crystals, formed by condensation of water vapor in the air. However, the base of a cloud is above the surface of the earth, while fog is in contact with the surface.

Radiation fog forms over low-lying land on clear, calm nights. As the land radiates heat and becomes cooler, it cools the air immedi- ately above the surface. This causes a **temperature inversion** to form, the temperature for some distance upward *increasing* with height. If the air is cooled to its dew point, fog forms. Often, cooler and more dense air drains down surrounding slopes to heighten the effect. Radiation fog is often quite shallow, and is usually thickest at the surface. After sunrise the fog may "lift," as shown in figure 7–12, and gradually dissipate, usually being entirely gone by noon. At sea

RADIATION FOG

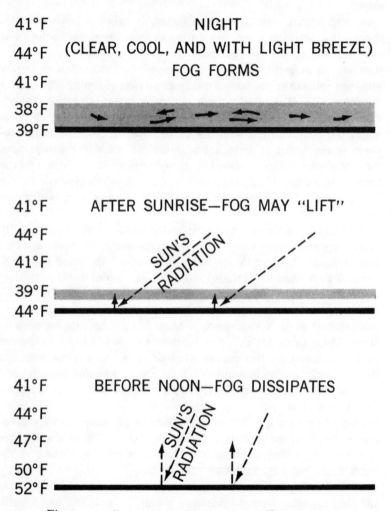

Figure 7–12 *Formation and dissipation of radiation fog.*

the temperature of the water undergoes little change between day and night, and so radiation fog is seldom encountered more than ten miles from shore.

Advection fog forms when warm, moist air blows over a colder surface and is cooled below its dew point. This type, most commonly encountered at sea, may be quite thick and often persists over relatively long periods. The maximum density might be at nearly any height. Advection fog is common over cold ocean currents. If the wind is strong enough to thoroughly mix the air, condensation may take place at some distance above the surface of the earth, forming low stratus clouds rather than fog.

Off the coast of California, winds create an offshore current which displaces the warm surface water, causing an upwelling of colder water. Moist air being transported along the coast in the same wind system is cooled, and advection fog results. In the coastal valleys, fog is sometimes formed when moist air blown inland during the afternoon is cooled by radiation during the night. Both of these are called **California fog** because they are peculiar to California and its coastal valleys.

When very cold air moves over warmer water, wisps of visible water vapor may rise from the surface as the water "steams." In extreme cases this **frost smoke,** or **arctic sea smoke,** may rise to a height of several hundred feet, the portion near the surface constituting a dense fog which obscures the horizon and surface objects, but usually leaves the sky relatively clear.

Fog consisting of ice crystals is called **ice fog,** or **pogonip** by Western American Indians. Thin fog of relatively large particles, or very fine rain lighter than drizzle, is called **mist.** A mixture of smoke and fog is called **smog.**

Haze consists of fine dust or salt particles in the air, too small to be individually apparent, but in sufficient number to reduce horizontal visibility and cast a bluish or yellowish veil over the landscape, subduing its colors and making objects appear indistinct. This is sometimes called **dry haze** to distinguish it from **damp haze,** which consists of small water droplets or moist particles in the air, smaller and more scattered than light fog. In international meteorological practice, the term "haze" is used to refer to a condition of atmospheric obscurity caused by dust and smoke.

Mirage—Light is refracted as it passes through the atmosphere. When refraction is normal, objects appear slightly elevated, and the visible horizon is farther from the observer than it otherwise would be. Since the effects are uniformly progressive, they are not apparent to the observer. When refraction is not normal, some form of **mirage** may occur. A mirage is an optical phenomenon in which objects appear distorted, displaced (raised or lowered), magnified, multiplied, or inverted due to varying atmospheric refraction which occurs when a layer of air near the earth's surface differs greatly in density from surrounding air. This may occur when there is a rapid and sometimes irregular change of temperature or humidity with height.

If there is a temperature inversion (increase of temperature with height), particularly if accompanied by a rapid decrease in humidity, the refraction is greater than normal. Objects appear elevated, and the visible horizon is farther away. Objects which are normally below the horizon become visible. This is called **looming.** If the upper portion of an object is raised much more than the bottom part, the object appears taller than usual, an effect called **towering.** If the lower part of an object is raised more than the upper part, the object appears shorter, an effect called **stooping.** When the refraction is greater than normal, a **superior mirage** may occur. An inverted image is seen above the object, and sometimes an erect image appears over the inverted one, with the bases of the two images touching. Greater than normal refraction usually occurs when the water is much colder than the air above it.

If the temperature decrease with height is much greater than normal, refraction is less than normal, or may even cause bending in the opposite direction. Objects appear lower than normal, and the visible horizon is closer to the observer. This is called **sinking.** Towering or stooping may occur if conditions are suitable. When the refraction is reversed, an **inferior mirage** may occur. A ship or an island appears to be floating in the air above a shimmering horizon, possibly with an inverted image beneath it. Conditions suitable to the formation of an inferior mirage occur when the surface is much warmer than the air above it. This usually requires a heated land mass, and therefore is more common near the coast than at sea.

When refraction is not uniformly progressive, objects may appear distorted, taking an almost endless variety of shapes. The sun when near the horizon is one of the objects most noticeably affected. A **fata morgana** is a complex mirage characterized by marked distortion, generally in the vertical. It may cause objects to appear towering, magnified, and at times even multiplied.

Sky Coloring—White light is composed of light of all colors. Color is related to wave length, the visible spectrum varying from about 0.000038 to 0.000076 centimeters. The characteristics of each color are related to its wave length (or frequency). Thus, the shorter the wave length, the greater the amount of bending when light is refracted. It is this principle that permits the separation of light from celestial bodies into a **spectrum** ranging from red, through orange, yellow, green, and blue, to violet, with long-wave infrared (black light) being slightly outside the visible range at one end and short-wave ultraviolet being slightly outside the visible range at the other end. Light of shorter wave length is scattered and diffracted more than that of longer wave length.

Light from the sun and moon is white, containing all colors. As it enters the earth's atmosphere, a certain amount of it is scattered. The blue and violet, being of shorter wave length than other colors, are scattered most. Most of the violet light is absorbed in the atmosphere. Thus, the scattered blue light is most apparent, and the sky appears blue. At great heights, above most of the atmosphere, it appears black.

When the sun is near the horizon, its light passes through more of the atmosphere than when higher in the sky, resulting in greater scattering and absorption of blue and green light, so that a larger percentage of the red and orange light penetrates to the observer. For this reason the sun and moon appear redder at this time, and when this light falls upon clouds, they appear colored. This accounts for the colors at sunset and sunrise. As the setting sun approaches the horizon, the sunset colors first appear as faint tints of yellow and orange. As the sun continues to set, the colors deepen. Contrasts occur, due principally to difference in height of clouds. As the sun sets, the clouds become a deeper red, first the lower clouds and then the higher ones, and finally they fade to a gray.

When there is a large quantity of smoke, dust, or other material in the sky, unusual effects may be observed. If the material in the atmosphere is of suitable substance and quantity to absorb the longer wave red, orange, and yellow radiations, the sky may have a greenish tint, and even the sun or moon may appear green. If the green light, too, is absorbed, the sun or moon may appear blue. A **green moon** or **blue moon** is most likely to occur when the sun is slightly below the horizon and the longer wave length light from the sun is absorbed, resulting in green or blue light being cast upon the atmosphere in front of the moon. The effect is most apparent if the moon is on the same side of the sky as the sun.

Rainbows—The familiar arc of concentric colored bands seen when the sun shines on rain, mist, spray, etc., is caused by refraction, internal reflection, and diffraction of sunlight by the drops of water. The center of the arc is a point 180° from the sun, in the direction of a line from the sun, through the observer. The radius of the brightest rainbow is 42°. The colors are visible because of the difference in the amount of refraction of the different colors making up white light, the light being spread out to form a spectrum. Red is on the outer side and blue and violet on the inner side, with orange, yellow, and green between, in that order from red.

Sometimes a secondary rainbow is seen outside the primary one, at a radius of about 50°. The order of colors of this rainbow is reversed. On rare occasions a faint rainbow is seen on the same side as the sun. The radius of this rainbow and the order of colors are the same as those of the primary rainbow.

A similar arc formed by light from the moon (a lunar rainbow) is called a **moonbow.** The colors are usually very faint. A faint, white arc of about 39° radius is occasionally seen in fog opposite the sun. This is called a **fogbow,** although its origin is controversial, some considering it a halo.

Halos—Refraction, or a combination of refraction and reflection, of light by ice crystals in the atmosphere (cirrostratus clouds) may cause a **halo** to appear. The most common form is a ring of light of radius 22° or 46° with the sun or moon at the center. Occasionally a faint, white circle with a radius of 90° appears around the sun. This is called a **Hevelian halo.** It is probably caused by refraction and internal reflection of the sun's light by bipyramidal ice crystals. A halo formed by refraction is usually faintly colored like a rainbow, with red nearest the celestial body, and blue farthest from it.

A brilliant rainbow-colored arc of about a quarter of a circle with its center at the zenith, and the bottom of the arc about 46° above the sun, is called a **circumzenithal arc.** Red is on the outside of the arc, nearest the sun. It is produced by the refraction and dispersion of the sun's light striking the top of prismatic ice crystals in the atmosphere. It usually lasts for only about five minutes, but may be so brilliant as to be mistaken for an unusually bright rainbow. A similar arc formed 46° *below* the sun, with red on the upper side, is called a **circumhorizontal arc.** Any arc tangent to a heliocentric halo (one surrounding the sun) is called a **tangent arc.** As the sun increases in elevation, such arcs tangent to the halo of 22° gradually bend their ends toward each other. If they meet, the elongated curve enclosing the circular halo is called a **circumscribed halo.** The inner edge is red.

A halo consisting of a faint, white circle through the sun and parallel to the horizon is called a **parhelic circle.** A similar one through the moon is called a **paraselenic circle.** They are produced by reflection of sunlight or moonlight from vertical faces of ice crystals.

A **parhelion** (plural *parhelia*) is a form of halo consisting of an image of the sun at the same altitude and some distance from it, usually 22°, but occasionally 46°. A similar phenomenon occurring at an angular distance of 120° (sometimes 90° or 140°) from the sun is called a **paranthelion.** One at an angular distance of 180°, a rare occurrence, is called an **anthelion,** although this term is also used to refer to a luminous, colored ring or **glory** sometimes seen around the shadow of one's head on a cloud or fog bank. A parhelion is popularly called a **mock sun** or **sun dog.** Similar phenomena in relation to the moon are called **paraselene** (popularly a **mock moon** or **moon dog**), **parantiselene,** and **antiselene.** The term *parhelion* should not be confused with *perihelion*, that orbital point nearest the sun when the sun is the center of attraction.

A **sun pillar** is a glittering shaft of white or reddish light occasionally seen extending above and below the sun, usually when the sun is near the horizon. A phenomenon similar to a sun pillar, but observed in connection with moon, is called a **moon pillar.** A rare form of halo in which horizontal and vertical shafts of light intersect at the sun is called a **sun cross.** It is probably due to the simultaneous occurrence of a sun pillar and a parhelic circle.

Corona—When the sun or moon is seen through altostratus clouds, its outline is indistinct, and it appears surrounded by a glow of light called a **corona.** This is somewhat similar in appearance to the corona seen around the sun during a solar eclipse. When the effect is due to clouds, however, the glow may be accompanied by one or more rainbow-colored rings of small radii, with the celestial body at the center. These can be distinguished from a halo by their much smaller radii and also by the fact that the order of the colors is reversed, red being on the inside, nearest the body, in the case of the halo, and on the outside, away from the body, in the case of the corona.

A corona is caused by diffraction of light by tiny droplets of water. The radius of a corona is inversely proportional to the size of the water droplets. A large corona indicates small droplets. If a corona decreases in size, the water droplets are becoming large and the air more humid. This may be an indication of an approaching rainstorm. The glow portion of a corona is called an **aureole.**

The green flash—As light from the sun passes through the atmosphere, it is refracted. Since the amount of bending is slightly different for each color, separate images of the sun are formed in each color of the spectrum. The effect is similar to that of imperfect color printing in which the various colors are slightly out of register. However, the difference is so slight that the effect is not usually noticeable. At the horizon, where refraction is maximum, the greatest difference, which occurs between violet at one end of the spectrum and red at the other, is about ten seconds of arc. At latitudes of the United States, about 0.7 second of time is needed for the sun to change altitude by this amount when it is near the horizon. The red image, being bent least by refraction, is first to set and last to rise. The shorter wave blue and violet colors are scattered most by the atmosphere, giving it its characteristic blue color. Thus, as the sun sets, the green image may be the last of the colored images to drop out of sight. If the red, orange, and yellow images are below the horizon, and the blue and violet light is scattered and absorbed, the upper rim of the green image is the only part seen, and the sun appears green. This is the **green flash.** The shade of green varies, and occasionally the blue image is seen, either separately or following the green flash (at sunset). On rare occasions the violet image is also seen. These colors may also be seen at sunrise, but in reverse order. They are occasionally seen when the sun disappears behind a cloud or other obstruction.

The phenomenon is not observed at each sunrise or sunset, but under suitable conditions is far more common than generally supposed. Conditions favorable to observation of the green flash are a sharp horizon, clear atmosphere, a temperature inversion, and an attentive observer. Since these conditions are more frequently met when the horizon is formed by the sea than by land, the phenomenon is more common at sea. With a sharp sea horizon and clear atmosphere, an attentive observer may see the green flash at as many as 50 percent of sunsets and sunrises, although a telescope may be needed for some of the observations.

Duration of the green flash (including the time of blue and violet flashes) of as long as ten seconds has been reported, but such length is rare. Usually it lasts for a period of about half a second to two and one-half seconds with about one and a quarter seconds being average. This variability is probably due primarily to changes in the index of refraction of the air near the horizon.

Under favorable conditions, a momentary green flash has been observed at the setting of Venus and Jupiter. A telescope improves the chances of seeing such a flash from a planet, but is not a necessity.

Crepuscular rays are beams of light from the sun passing through openings in the clouds, and made visible by illumination of dust in the atmosphere along their paths. Actually, the rays are virtually parallel, but because of perspective appear to diverge. Those appearing to extend downward are popularly called **backstays of the sun,** or **sun drawing water.** Those extending upward and across the sky, appearing to converge toward a point 180° from the sun, are called **anticrepuscular rays.**

The Atmosphere and Radio Waves—Radio waves traveling through the atmosphere exhibit many of the properties of light, being refracted, reflected, diffracted, and scattered.

Atmospheric Electricity—Various conditions induce the formation of electrical charges in the atmosphere. When this occurs, there is often a difference of electron charge between various parts of the atmosphere, and between the atmosphere and earth or terrestrial objects. When this difference exceeds a certain minimum value depending upon the conditions, the static electricity is discharged, resulting in phenomena such as lightning or St. Elmo's fire.

Lightning is the discharge of electricity from one part of a thundercloud to another, from one such cloud to another, or between such a cloud and the earth or a terrestrial object.

Enormous electrical stresses build up within thunderclouds and between such clouds and the earth. At some point the resistance of the intervening air is overcome. At first the process is a progressive one, probably starting as a brush discharge (St. Elmo's fire) and growing by ionization. The breakdown follows an irregular path along the line of least resistance. A hundred or more individual discharges may be necessary to complete the path between points of opposite polarity. When this "leader stroke" reaches its destination, a heavy "main stroke" immediately follows in the opposite direction. This main stroke is the visible lightning, which may be tinted any color, depending upon the nature of the gases through which it passes. The illumination is due to the high degree of ionization of the air, which causes many of the atoms to be in excited states and emit radiation.

Thunder, the noise that often accompanies lightning, is caused by the heating and ionizing of the air by lightning, which results in rapid expansion of the air along its path and the sending out of a compression wave. Thunder may be heard at a distance of as much as 15 miles, but generally does not carry that far. The elapsed time between the flash of lightning and reception of the accompanying sound of thunder is an indication of the distance, because of the difference in travel time of light and sound. Since the former is comparatively instantaneous, and the speed of sound is about 1,117 feet per second, the approximate distance in nautical miles is equal to the elapsed time in seconds, divided by 5.5. If there is no accompanying thunder, the flash is called **heat lightning.**

Figure 7–13 *Waterspouts.*

St. Elmo's fire is a luminous discharge of electricity from pointed objects such as the masts and yardarms of ships, lightning rods, steeples, mountain tops, blades of grass, human hair, arms, etc., when there is a considerable difference in the electrical charge between the object and the air. It appears most frequently during a storm. An object from which St. Elmo's fire emanates is in danger of being struck by lightning, since this type discharge may be the initial phase of the leader stroke. Throughout history those who have not understood St. Elmo's fire have regarded it with superstitious awe, considering it a supernatural manifestation. This view is reflected in the name **corposant** (from "corpo santo," meaning "body of a saint"), sometimes given this phenomenon.

The **aurora** is a luminous glow appearing in varied forms in the thin atmosphere high above the earth, due to radiation from the sun.

Waterspouts—A waterspout is a small, whirling storm over the ocean or inland waters. Its chief characteristic is a funnel-shaped cloud extending, in a fully developed spout, from the surface of the water to the base of a cumulus type cloud (figure 7–13). The water in a spout is mostly confined to its lower portion, and may be either salt spray drawn up by the sea surface or fresh water resulting from condensation due to the lowered pressure in the center of the vortex creating the spout. Waterspouts usually rotate in the same direction as cyclones (counterclockwise in the northern hemisphere and clockwise in the southern hemisphere), but the opposite rotation is occasionally observed. They are found most frequently in tropical regions, but are not uncommon in higher latitudes.

Waterspouts may be divided into two classes, according to their different origins and appearances. In the true waterspout, the vortex is formed in clouds by the interaction of air currents flowing in opposite directions. This type occurs mainly in the vicinity of a squall line. A similar disturbance over land is called a **tornado.** The second type, which may be considered a pseudo waterspout, originates just above the water surface, in unstable air, and builds upward, frequently under clear skies. This type is identical to the whirling pillars of sand and dust often seen on deserts and usually occurs only over very warm water surfaces.

Waterspouts vary in diameter from a few feet to several hundred feet, and in height from a few hundred feet to several thousand feet. Sometimes they assume fantastic shapes and may even seem to coil about themselves. Since a waterspout is often inclined to the vertical, its actual length may be much greater than indicated by its height.

Forecasting Weather—The prediction of weather at some future time is based upon an understanding of weather processes, and observations of present conditions. Thus, one learns that when there is a certain sequence of cloud types, rain can usually be expected to follow within a certain period. If the sky is cloudless, more heat will be received from the sun by day, and more heat will be radiated outward from the warm earth by night than if the sky is overcast. If the wind is in such a direction that warm, moist air will be transported to a colder surface, fog can be expected. A falling barometer indicates the approach of a "low," probably accompanied by stormy weather. Thus, before the science of meteorology was developed, many individuals learned to interpret certain phenomena in terms of future weather, and to make reasonably accurate forecasts for short periods into the future.

With the establishment of weather observation stations, additional information became available. As such observations expanded, and communication facilities improved, knowledge of simultaneous conditions over wider areas became available. This made possible the collection of these "synoptic" reports at civilian forecast centers and Navy Fleet Weather Centrals.

The individual observations are made at government-operated

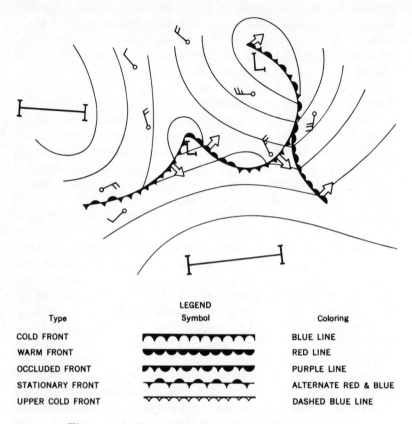

LEGEND

Type	Symbol	Coloring
COLD FRONT		BLUE LINE
WARM FRONT		RED LINE
OCCLUDED FRONT		PURPLE LINE
STATIONARY FRONT		ALTERNATE RED & BLUE
UPPER COLD FRONT		DASHED BLUE LINE

Figure 7–14 *Designation of fronts on weather maps.*

stations on shore, and aboard vessels at sea. Observations aboard merchant ships at sea are made and transmitted on a voluntary and cooperative basis. The various national meteorological services supply shipmasters with blank forms, printed instructions, and other materials essential to the making, recording, and interpreting of observations. Shipmasters render a valuable service by reporting all contacts with tropical cyclones.

Symbols and numbers are used to indicate on a **synoptic chart,** popularly called a **weather map,** the conditions at each observation station. Isobars are drawn through lines of equal atmospheric pressure, fronts are located and marked by symbol (figure 7–14) areas of precipitation and fog are indicated, etc.

Ordinarily, surface charts are prepared every six hours, but at a few centers they are drawn every three hours. In addition, synoptic charts for selected heights are prepared two to four times per day. Knowledge of conditions aloft is of value in establishing the three-dimensional structure of the atmosphere at any time, and the motions upon which forecasts are based.

By studying the latest synoptic weather chart and comparing it with previous charts, a trained meteorologist having a knowledge of local weather peculiarities can draw certain inferences regarding future weather, and issue a forecast. Weather forecasts are essentially a form of extrapolation. Past changes and present trends are used to predict future events. In areas where certain sequences follow with great certainty, the probability of an accurate forecast is very high. In transitional areas, or areas where an inadequate number of synoptic reports is available, the forecasts are less reliable. Forecasts, then, are based upon the principles of probability, and where nature provides low probability, high reliability should not be expected. In any area, the probability of a given event occurring decreases with the lead time. Thus, a forecast for six hours after a synoptic chart is drawn should be more reliable than one for 24 hours ahead. Long-term forecasts for two weeks or a month in advance are limited to general statements. For example, a prediction is made as to which areas will have temperatures above or below normal, and how precipitation will compare with normal, but no attempt is made to state that rainfall will occur at a certain time and place.

Forecasts are issued for various areas. The national meteorological services of most maritime nations, including the United States, issue forecasts for ocean areas and warnings of the approach of storms. The efforts of the various nations are coordinated through the World Meteorological Organization.

Dissemination of weather information is carried out in a number of ways. Forecasts are widely broadcast by commercial and government radio stations, and printed in newspapers. Visual storm warnings are displayed in various ports, and storm warnings are broadcast by radio. (See Appendix B.)

Through the use of codes, a simplified version of synoptic weather charts is transmitted to various stations ashore and afloat. Rapid transmission of completed maps has been made possible by the development of facsimile transmitters and receivers. This system is based upon detailed scanning, by a photoelectric detector, of properly illuminated black and white copy. The varying degrees of light intensity are converted to electric energy which is transmitted to the receiver and converted back to a black and white presentation.

Complete information on dissemination of weather information by radio is given in H.O. Pubs. Nos. 118–A and 118–B, *Radio Weather Aids*. This publication lists broadcast schedules and weather codes. Information on day and night visual storm warnings is given in the various volumes of sailing directions and coast pilots.

Interpreting the weather—The factors which determine weather are numerous and varied. Ever-increasing knowledge regarding them makes possible a continually improving weather service. However, the ability to forecast is acquired through study and long practice, and therefore the services of a trained meteorologist should be utilized whenever available.

The value of a forecast is increased if one has access to the information upon which it is based, and understands the principles and processes involved. It is sometimes as important to know the various types of weather that *might* be experienced as it is to know which of several possibilities is *most likely* to occur.

At sea, reporting stations are unevenly distributed, sometimes leaving relatively large areas with incomplete reports, or none at all. Under these conditions, the locations of highs, lows, fronts, etc., are imperfectly known, and their very existence may even be in doubt. At such times the mariner who can interpret the observations made from his own vessel may be able to predict weather during the next 24 hours more reliably than a trained meteorologist some distance away with incomplete information.

Influencing the Weather—Meteorological activities are devoted primarily to understanding weather processes, and predicting future weather. As knowledge regarding cause-and-effect relationships increases, the possibility of being able to induce certain results by artificially producing the necessary conditions becomes greater. The most promising results to date have been in the encouraging of precipitation by "seeding" supercooled clouds with powdered dry ice or silver iodide smoke. The effectiveness of this procedure is controversial. Various methods of decreasing the intensity of tropical cyclones, or of diverting their courses, have been suggested, but a satisfactory method has not been devised.

If a way is found to influence weather on a major scale, legal and possibly moral problems will be created due to conflicting interests.

=8=

Weather Observations

Weather forecasts are generally based upon information acquired by observations made at a large number of stations. Ashore, these stations are located so as to provide adequate coverage of the area of interest. Most observations at sea are made by mariners, wherever they happen to be. Since the number of observations at sea is small compared to the number ashore, marine observations are of importance in areas where little or no information is available from other sources. Results of these observations are recorded in the deck log, or other appropriate form. Data recorded by designated vessels are sent by radio to centers ashore, where they are plotted, along with other observations, to provide data for drawing synoptic charts. These charts are used to make forecasts. Complete weather information gathered at sea is mailed to the appropriate meteorological services for use in the preparation of weather atlases and in marine climatological studies.

The analysis of the weather map can be no better than the weather reports used for making the map. A knowledge of weather elements and the instruments used to measure them is therefore of importance to the mariner who hopes to benefit from weather forecasts.

Instruments of various types have been developed to aid in making weather observations. Some have been in use for many years, while others have been developed only recently. Electronic devices have aided materially, but the full impact of electronics upon meteorology has not yet been felt. Several new types of electronic weather instruments are in various stages of development.

Atmospheric Pressure Measurement—The sea of air surrounding the earth exerts a pressure of about 14.7 pounds per square inch on the surface of the earth. This **atmospheric pressure,** sometimes called **barometric pressure,** varies from place to place, and at the same place it varies with time.

Atmospheric pressure is one of the basic elements of a meteorological observation. When the pressure at each station is plotted on a synoptic chart, lines of equal atmospheric pressure, called **isobars,** are drawn to indicate the areas of high and low pressure and their centers. These are useful in making weather predictions, because certain types of weather are characteristic of each type area, and often the wind patterns over large areas are deduced from the isobars.

Atmospheric pressure is measured by means of a **barometer.** A **mercurial barometer** does this by balancing the weight of a column of air against that of a column of mercury. The **aneroid barometer** has a partly evacuated, thin-metal cell which is compressed by atmospheric pressure, the amount of the compression being related to the pressure.

Early mercurial barometers were calibrated to indicate the height, usually in inches or millimeters, of the column of mercury needed to balance the column of air above the point of measurement. While the units **inches of mercury** and **millimeters of mercury** are still widely used, many modern barometers are calibrated to indicate the centimeter-gram-second unit of pressure, the **millibar,** which is equal to 1,000 dynes per square centimeter. A **dyne** is the force required to accelerate a mass of one gram at the rate of one centimeter per second per second.

The mercurial barometer was invented by Evangelista Torricelli in 1643. In its simplest form it consists of a glass tube a little more than 30 inches in length and of uniform internal diameter; one end being closed, the tube is filled with mercury, and inverted into a cup of mercury. The mercury in the tube falls until the column is just supported by the pressure of the atmosphere on the open cup, leaving a vacuum at the upper end of the tube. The height of the column indicates atmospheric pressure, greater pressures supporting higher columns of mercury.

The mercurial barometer is subject to rapid variations in height, called **pumping,** due to pitch and roll of the vessel and temporary changes in atmospheric pressure in the vicinity of the barometer. Because of this, the care required in the reading of the instrument, its bulkiness, and its vulnerability to physical damage, the mercurial barometer has been largely replaced at sea by the aneroid barometer.

The aneroid barometer (figure 8–1) measures atmospheric pressure by means of the force exerted by the pressure on a partly evacuated, thin-metal element called a **sylphon cell.** A small spring is used, either internally or externally, to partly counteract the tendency of

Figure 8–1 *An aneroid barometer.*

the atmospheric pressure to crush the cell. Atmospheric pressure is indicated directly by a scale and a pointer connected to the cell by a combination of levers. The linkage provides considerable magnification of the slight motion of the cell, to permit readings to higher precision than could be obtained without it.

An aneroid barometer should be mounted permanently. Prior to installation, the barometer should be carefully set to station pressure. An adjustment screw is provided for this purpose. The error in the reading of the instrument is determined by comparison with a mercurial barometer or a standard precision aneroid barometer. If a qualified meteorologist is not available to make this adjustment, it is good practice to remove only one-half the apparent error. The case should then be tapped gently to assist the linkage to adjust itself, and the process repeated. If the remaining error is not more than half a millibar (0.015 inch), no attempt should be made to remove it by further adjustment. Instead, a correction should be applied to the readings. The accuracy of this correction should be checked from time to time.

A **precision aneroid barometer** used at weather stations ashore, and for comparison of shipboard instruments, is constructed and

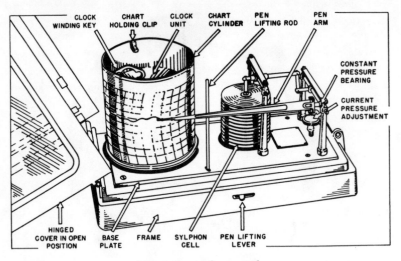

Figure 8–2 *A barograph.*

tested to more exacting tolerances than the ordinary barometer, and provides readings to greater accuracy.

The barograph (figure 8–2) is a recording barometer. Basically, it is the same as a nonrecording aneroid barometer except that the pointer carries a pen at its outer end, and the scale is replaced by a slowly rotating cylinder around which a prepared chart has been wrapped. A clock mechanism inside the cylinder rotates the cylinder so that a continuous line is traced on the chart, recording the pressure at any time.

A microbarograph is a precise barograph with greater magnification of deformations due to pressure changes, and a correspondingly expanded chart. Two sylphon cells are used, one being mounted over the other in tandem. Minor fluctuations due to shocks or vibrations are eliminated by damping. Since oil-filled dashpots are used for this purpose, the instrument should not be inverted.

The barograph is usually mounted on a shelf or desk in a room open to the atmosphere, and in a location which minimizes the effect of the ship's vibration. Shock-absorbing material such as sponge rubber is placed under the instrument to minimize the transmission of shocks.

The pen should be checked and the inkwell filled each time the chart is changed, every week in the case of the barograph, and every four days in the case of the microbarograph. The dashpots of the micro-barograph should be kept filled with dashpot oil to within three-eighths inch of the top.

Both instruments require checking from time to time to insure correct indication of pressure. The position of the pen is adjusted by a small knob provided for this purpose. The adjustment should be made in stages, eliminating half the apparent error, tapping the case to insure linkage adjustment to the new setting, and then repeating the process.

Adjustment of Barometer Readings—Atmospheric pressure as indicated by a barometer or barograph may be subject to several errors, as follows:

Instrument Error—Any inaccuracy due to imperfection or incorrect adjustment of the instrument can be determined by comparison with a standard instrument. The National Weather Service provides a comparison service. In certain ports a representative brings a standard barometer on board ships for comparison, or a barometer can be taken to a local National Weather Service office, and comparison can be made there. The correct sea-level pressure can be obtained by telephone. The shipboard barometer should be corrected for height, as explained below, before comparison with this telephoned value. If there is reason to believe that the barometer is in error, it should be compared with a standard, and if an error is found, the barometer should be adjusted to the correct reading, or a correction applied to all readings.

Height Error—Since atmospheric pressure is caused by the weight of air above the place, the pressure decreases as height increases. The correct value at the barometer is called **station pressure.** Isobars adequately reflect wind conditions and geographic distribution of pressure only when they are drawn for pressure at constant height (or the varying height at which a constant pressure exists). On synoptic charts it is customary to show the equivalent pressure at sea level, called **sea level pressure.** This is found by applying a correction to station pressure. The correction, given in Table 8–1, depends upon the height of the barometer and the average temperature of the air between this height and the surface. The outside air temperature taken aboard ship is sufficiently accurate for this purpose. *This is an important correction which should be applied to all readings of any type barometer.*

Gravity Error—Mercurial barometers are calibrated for standard sea level gravity at latitude 45°32′40″. If the gravity differs from this

TABLE 8–1

Correction of Barometer Reading for Height Above Sea Level

All barometers. All values positive.

Height in feet	Outside temperature in degrees Fahrenheit													Height in feet
	−20°	−10°	0°	10°	20°	30°	40°	50°	60°	70°	80°	90°	100°	
	Inches	*Inches*	*Inches*	*Inches*	*Inches*	*Inches*	*Inches*	*Inches*	*Inches*	*Inches*	*Inches*	*Inches*	*Inches*	
5	0.01	0.01	0.01	0.01	0.01	0.01	0.01	0.01	0.01	0.01	0.01	0.01	0.01	5
10	0.01	0.01	0.01	0.01	0.01	0.01	0.01	0.01	0.01	0.01	0.01	0.01	0.01	10
15	0.02	0.02	0.02	0.02	0.02	0.02	0.02	0.02	0.02	0.02	0.02	0.02	0.02	15
20	0.03	0.02	0.02	0.02	0.02	0.02	0.02	0.02	0.02	0.02	0.02	0.02	0.02	20
25	0.03	0.03	0.03	0.03	0.03	0.03	0.03	0.03	0.03	0.03	0.03	0.03	0.03	25
30	0.04	0.04	0.04	0.04	0.04	0.03	0.03	0.03	0.03	0.03	0.03	0.03	0.03	30
35	0.04	0.04	0.04	0.04	0.04	0.04	0.04	0.04	0.04	0.04	0.04	0.04	0.04	35
40	0.05	0.05	0.05	0.05	0.05	0.05	0.04	0.04	0.04	0.04	0.04	0.04	0.04	40
45	0.06	0.06	0.05	0.05	0.05	0.05	0.05	0.05	0.05	0.05	0.05	0.05	0.05	45
50	0.06.	0.06	0.06	0.06	0.06	0.06	0.06	0.06	0.06	0.05	0.05	0.05	0.05	50
55	0.07	0.07	0.07	0.07	0.06	0.06	0.06	0.06	0.06	0.06	0.06	0.06	0.06	55
60	0.08	0.07	0.07	0.07	0.07	0.07	0.07	0.07	0.06	0.06	0.06	0.06	0.06	60
65	0.08	0.08	0.08	0.08	0.08	0.07	0.07	0.07	0.07	0.07	0.07	0.07	0.07	65
70	0.09	0.09	0.09	0.08	0.08	0.08	0.08	0.08	0.08	0.08	0.07	0.07	0.07	70
75	0.10	0.09	0.09	0.09	0.09	0.09	0.09	0.08	0.08	0.08	0.08	0.08	0.08	75
80	0.10	0.10	0.10	0.10	0.09	0.09	0.09	0.09	0.09	0.09	0.08	0.08	0.08	80
85	0.11	0.11	0.10	0.10	0.10	0.10	0.10	0.09	0.09	0.09	0.09	0.09	0.09	85
90	0.11	0.11	0.11	0.11	0.11	0.10	0.10	0.10	0.10	0.10	0.09	0.09	0.09	90
95	0.12	0.12	0.12	0.11	0.11	0.11	0.11	0.10	0.10	0.10	0.10	0.10	0.10	95
100	0.13	0.12	0.12	0.12	0.12	0.11	0.11	0.11	0.11	0.11	0.10	0.10	0.10	100
105	0.13	0.13	0.13	0.13	0.12	0.12	0.12	0.12	0.11	0.11	0.11	0.11	0.11	105
110	0.14	0.14	0.13	0.13	0.13	0.13	0.12	0.12	0.12	0.12	0.11	0.11	0.11	110
115	0.15	0.14	0.14	0.14	0.13	0.13	0.13	0.13	0.12	0.12	0.12	0.12	0.12	115
120	0.15	0.15	0.15	0.14	0.14	0.14	0.13	0.13	0.13	0.13	0.12	0.12	0.12	120
125	0.16	0.16	0.15	0.15	0.15	0.14	0.14	0.14	0.13	0.13	0.13	0.13	0.12	125

amount, an error is introduced and must be corrected. *This correction does not apply to readings of an aneroid barometer.* Gravity also changes with height above sea level, but the effect is negligible for the first few hundred feet, and so is not needed for readings taken aboard ship.

Temperature Error—Barometers are calibrated at a standard temperature of 32° F. The liquid of a mercurial barometer expands as the temperature of the mercury rises, and contracts as it decreases. Modern aneroid barometers are compensated for temperature changes by the use of different metals having unequal coefficients of linear expansion.

Wind measurement consists of determination of the direction *from* which the wind is blowing, and the speed of the wind. Wind direction is measured by a **wind vane,** and wind speed by an **anemometer.**

A wind vane consists of a device pivoted on a vertical shaft, with more surface area on one side of the pivot than on the other, so that the wind exerts more force on one side, causing the smaller end to point into the wind. An indicator may be connected to the shaft to provide continuous measurement of wind direction.

In its simplest form, an anemometer consists of a number of cups mounted on short horizontal arms attached to a longer vertical shaft which rotates as the wind blows against the cups. The speed at which the shaft rotates is directly proportional to the wind speed. The number of rotations may be indicated by a counter or by marks on a revolving drum, or the speed may be indicated directly by a device similar to an automobile speedometer. Still another method is to connect a buzzer or flashing light so calibrated that the number of signals per unit time is the speed in knots or miles per hour.

The standard anemometer used aboard ship has three cups. Some anemometers have four cups, and certain naval vessels use a type called the **bridled cup anemometer,** which has a large number of cups mounted on a shaft which does not rotate freely. An anemometer which uses a propeller as the rotor to measure wind speed, and has a streamlined, tail-type vane to indicate direction, is being installed on some ships. Similar equipment is used ashore, customarily mounted on a guyed mast 13 feet high. Wind direction is transmitted to an indicator or recorder by a synchronous motor, while wind speed is transmitted as a voltage generated by a direct-current magneto driven by the propeller. A synchro system is connected to some wind-measuring equipment to provide remote indication of the velocity (both direction and speed). Lightweight, portable, hand-held instruments for measuring and indicating wind speed in knots are used on some ships, principally aircraft carriers.

Several types of wind speed and direction **recorders** are available. Each instrument is normally supplied with a description and complete operating instructions.

If no anemometer is available, wind speed can be estimated by its effect upon the sea and objects in its path, as explained below.

True and Apparent Wind—An observer aboard a vessel proceeding through still air experiences an **apparent wind** which is from dead ahead and has an apparent speed equal to the speed of the vessel. Thus, if the actual or **true wind** is zero and the speed of the vessel is ten knots, the apparent wind is from dead ahead at ten knots. If the true wind is from dead ahead at 15 knots, and the speed of the vessel is ten knots, the apparent wind is $15 + 10 = 25$ knots from dead ahead. If the vessel makes a $180°$ turn, the apparent wind is $15 - 10 = 5$ knots from dead astern.

In any case, the apparent wind is the vector sum of the true wind and the *reciprocal* of the vessel's course and speed vector. Since wind vanes and anemometers measure *apparent* wind, the usual problem aboard a vessel equipped with an anemometer is to convert this to true wind. There are several ways of doing this. Perhaps the simplest is by the graphical solution illustrated in the following example:

Example 1—A ship is proceeding on course 150° at a speed of 17 knots. The apparent wind is from 40° off the starboard bow, speed 15 knots.

Required—The relative direction, true direction, and speed of the true wind.

Solution (figure 8–3)—Starting at the center of a maneuvering board or other suitable form, draw a line in the relative direction *from* which the apparent wind is blowing. Locate point 1 on this line, at a distance from the center equal to the speed of the apparent wind (2:1 scale is used in figure 8–3). From point 1, draw a line vertically *downward*. Locate point 2 on this line at a distance from point 1 equal to the speed of the vessel in knots, to the same scale as the first line. The relative direction of the true wind is *from* point 2 (120°) toward the center, and the speed of the true wind is the distance of point 2 from the center, to the same scale used previously (11 kn.). The true direction of the wind is the relative direction plus the true heading, or 120° + 150° = 270°.

Answers—True wind from 120° relative, 270° true, at 11 knots.

A quick solution can be made without an actual plot, in the following manner: On a maneuvering board (H.O. 2665–10), label the circles 5, 10, 15, 20, etc., from the center, and draw vertical lines tangent to these circles. Cut out the 5:1 scale and discard that part having graduations greater than the maximum speed of the vessel. Keep this equipment for all solutions. (For durability, the two parts can be mounted on cardboard or other suitable material.) To find true wind, spot in point 1 by eye. Place the zero of the 5:1 scale on this point and align the scale (inverted) by means of the vertical lines. Locate point 2 at the speed of the vessel as indicated on the 5:1 scale. It is always vertically *below* point 1. Read the relative direction and the speed of the true wind using eye interpolation if needed. The U.S. Weather Bureau distributes a wind vector computer called a *Shipboard Wind Plotter* (figure 8–4). Solution by means of this plotter is illustrated in the following example:

Example 2—A ship is proceeding on course 270° at a speed of 14.5 knots. The apparent wind is from 40° off the starboard bow, speed 20 knots.

Required—The relative direction, true direction, and speed of the true wind by U.S. Weather Bureau Shipboard Wind Plotter.

Solution (figure 8–4)—The true direction of the apparent wind is

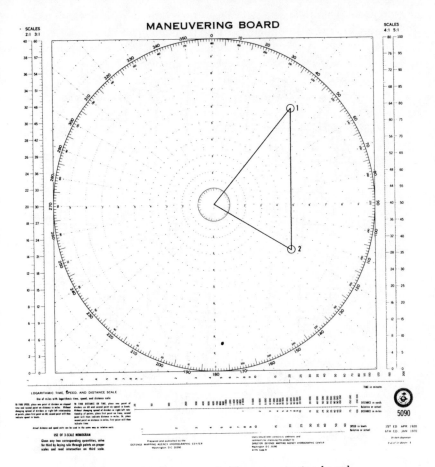

Figure 8–3 *Finding true wind by maneuvering board.*

determined by adding the apparent wind direction to the ship's heading if the wind is from off the starboard bow and subtracting the apparent wind direction if the wind is from off the port bow. In this example, the true direction of the apparent wind is 310°. In this solution the red arrowhead is considered the top of the plotter. Set ship's course, 270°, to the top of the plotter by rotating the protractor disk to set 270° at the red arrow. Using a convenient linear scale, measure vertically downward from the center peg of the plotting board a distance equivalent to 14.5 knots. Mark this point "S" for ship. Rotate the protractor disk of the plotting board until 310° is at

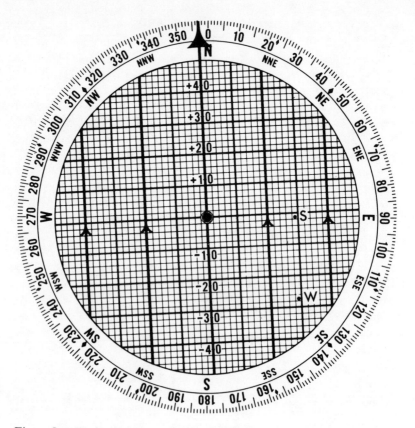

Figure 8–4 *Finding true wind by Weather Bureau Shipboard Wind Plotter.*

the red arrowhead at the top of the plotting board. Using the same linear scale as for ship's speed, plot vertically downward from the center peg of the plotting board a distance equivalent to 20 knots. Mark this point "W". Rotate the protractor disk until the "S" is vertically above the "W", using the vertical lines on the plotting board to line up the two points. Read the true wind direction at the top of the plotting board. The distance between points "S" and "W" is the true wind speed, using the same scale as in plotting points "S" and "W".

Answers—True wind direction is 357°, true wind speed is 13 knots.

Wind speed determined by appearance of the sea is the speed of the true wind. The sea also provides an indication of the direction of the true wind, because waves move in the same direction as the generating wind. If a wind vane is used, the direction of the apparent wind thus determined can be used with the speed of the true wind to determine the direction of the true wind by vector diagram. If a maneuvering board is used, draw a circle about the center equal to the speed of the true wind. From the center, plot the ship's vector (true course and speed). From the end of this vector draw a line in the direction in which the apparent wind is blowing (reciprocal of the direction from which it is blowing) until it intersects the speed circle. This line is the apparent wind vector, its length denotes the speed. A line from the center of the board to the end of the apparent wind vector is the true wind vector. The reciprocal of this vector is the direction from which the true wind is blowing. If the true wind speed is less than the speed of the vessel, two solutions are possible. If solution is by Table 8–2, the true speed, in units of ship's speed, is found in the column for the direction of the apparent wind. The number to the left is the relative direction of the true wind. The number on the same line in the side columns is the speed of the apparent wind in units of ship's speed. Again, two solutions are possible if true wind speed is less than ship's speed.

Wind and the Sea—There is a relationship between the speed of the wind and the state of the sea in the immediate vicinity of the wind. This is useful in predicting the sea conditions to be anticipated when future wind speed forecasts are available. It can also be used to estimate the speed of the wind, which may be desirable when an anemometer is not available.

Wind speeds are usually grouped in accordance with the **Beaufort scale,** (Appendix C), named after Admiral Sir Francis Beaufort, who devised it in 1806. As adopted in 1838, Beaufort numbers ranged from 0, calm, to 12, hurricane. They have now been extended to 17. The wind speed and the appearance of the sea at Beaufort scale numbers from 0 through 12 are shown in figures 8–5 through 8–17.

Temperature—Temperature is the intensity or degree of heat. It is measured in degrees. Several different temperature scales are in use.

On the **Fahrenheit** (**F**) scale commonly used in the United States and other English-speaking countries, pure water freezes at $32°$ and boils at $212°$.

On the **Celsius** (**C**) scale commonly used with the metric system, the freezing point of pure water is $0°$ and the boiling point is $100°$. This scale has been known by various names in different countries.

TABLE 8-2
Direction and Speed of True Wind in Units of Ship's Speed

Apparent wind speed	0°		10°		20°		30°		40°		Apparent wind speed
0.0	180	1.00	180	1.00	180	1.00	180	1.00	180	1.00	0.0
0.1	180	0.90	179	0.90	178	0.91	177	0.91	176	0.93	0.1
0.2	180	0.80	178	0.80	175	0.81	173	0.83	171	0.86	0.2
0.3	180	0.70	176	0.71	172	0.73	169	0.76	166	0.79	0.3
0.4	180	0.60	173	0.61	168	0.64	163	0.68	160	0.74	0.4
0.5	180	0.50	170	0.51	162	0.56	156	0.62	152	0.70	0.5
0.6	180	0.40	166	0.42	155	0.48	148	0.57	144	0.66	0.6
0.7	180	0.30	159	0.33	145	0.42	138	0.53	136	0.65	0.7
0.8	180	0.20	147	0.25	132	0.37	127	0.50	127	0.64	0.8
0.9	180	0.10	126	0.19	117	0.34	116	0.50	118	0.66	0.9
1.0	calm	0.00	95	0.17	100	0.35	105	0.52	110	0.68	1.0
1.1	0	0.10	66	0.21	85	0.38	95	0.55	103	0.72	1.1
1.2	0	0.20	49	0.28	73	0.43	86	0.60	96	0.78	1.2
1.3	0	0.30	39	0.36	64	0.50	79	0.66	90	0.84	1.3
1.4	0	0.40	33	0.45	57	0.57	73	0.73	85	0.90	1.4
1.5	0	0.50	29	0.54	51	0.66	68	0.81	81	0.98	1.5
1.6	0	0.60	26	0.64	47	0.74	64	0.89	78	1.05	1.6
1.7	0	0.70	24	0.74	44	0.83	61	0.97	75	1.13	1.7
1.8	0	0.80	22	0.83	42	0.93	58	1.06	72	1.22	1.8
1.9	0	0.90	21	0.93	40	1.02	56	1.15	70	1.30	1.9
2.0	0	1.00	20	1.03	38	1.11	54	1.24	68	1.39	2.0
2.5	0	1.50	17	1.52	32	1.60	47	1.71	60	1.85	2.5
3.0	0	2.00	15	2.02	29	2.09	43	2.19	56	2.32	3.0
3.5	0	2.50	14	2.52	28	2.58	41	2.68	53	2.81	3.5
4.0	0	3.00	13	3.02	26	3.08	39	3.17	51	3.30	4.0
4.5	0	3.50	13	3.52	25	3.58	38	3.67	50	3.79	4.5
5.0	0	4.00	12	4.02	25	4.08	37	4.16	49	4.28	5.0
6.0	0	5.00	12	5.02	24	5.07	36	5.16	47	5.27	6.0
7.0	0	6.00	12	6.02	23	6.07	35	6.15	46	6.27	7.0
8.0	0	7.00	11	7.02	23	7.07	34	7.15	45	7.26	8.0
9.0	0	8.00	11	8.02	22	8.07	34	8.15	44	8.26	9.0
10.0	0	9.00	11	9.02	22	9.06	33	9.15	44	9.26	10.0

	50°		60°		70°		80°		90°		
0.0	180	1.00	180	1.00	180	1.00	180	1.00	180	1.00	0.0
0.1	175	0.94	175	0.95	174	0.97	174	0.99	174	1.00	0.1
0.2	170	0.88	169	0.92	169	0.95	168	0.99	169	1.02	0.2
0.3	164	0.84	163	0.89	163	0.94	163	0.99	163	1.04	0.3
0.4	158	0.80	157	0.87	156	0.94	157	1.01	158	1.08	0.4
0.5	151	0.78	150	0.87	150	0.95	152	1.04	153	1.12	0.5
0.6	143	0.77	143	0.87	145	0.97	147	1.07	149	1.17	0.6
0.7	136	0.77	137	0.89	139	1.01	142	1.12	145	1.22	0.7
0.8	128	0.78	131	0.92	134	1.05	138	1.17	141	1.28	0.8
0.9	121	0.81	125	0.95	129	1.09	134	1.22	138	1.35	0.9
1.0	115	0.85	120	1.00	125	1.15	130	1.29	135	1.41	1.0
1.1	109	0.89	115	1.05	121	1.21	127	1.35	132	1.49	1.1
1.2	104	0.95	111	1.11	118	1.27	124	1.42	130	1.56	1.2
1.3	99	1.01	107	1.18	114	1.34	121	1.50	128	1.64	1.3
1.4	95	1.08	104	1.25	112	1.42	119	1.57	126	1.72	1.4
1.5	92	1.15	101	1.32	109	1.49	117	1.65	124	1.80	1.5
1.6	89	1.23	98	1.40	107	1.57	115	1.73	122	1.89	1.6
1.7	86	1.31	96	1.48	105	1.65	113	1.82	120	1.97	1.7
1.8	84	1.39	94	1.56	103	1.73	111	1.90	119	2.06	1.8
1.9	81	1.47	92	1.65	101	1.82	110	1.99	118	2.15	1.9
2.0	79	1.56	90	1.73	100	1.91	108	2.07	117	2.24	2.0
2.5	72	2.01	83	2.18	94	2.35	103	2.53	112	2.69	2.5
3.0	68	2.48	79	2.65	89	2.82	99	2.99	108	3.16	3.0
3.5	65	2.96	76	3.12	87	3.29	96	3.47	106	3.64	3.5
4.0	63	3.44	74	3.61	84	3.78	94	3.95	104	4.12	4.0
4.5	61	3.93	72	4.09	83	4.26	93	4.44	103	4.61	4.5
5.0	60	4.42	71	4.58	81	4.75	92	4.93	101	5.10	5.0
6.0	58	5.41	69	5.57	79	5.74	90	5.91	99	6.08	6.0
7.0	57	6.40	68	6.56	78	6.72	88	6.90	98	7.07	7.0
8.0	56	7.40	67	7.55	77	7.72	87	7.89	97	8.06	8.0
9.0	55	8.39	66	8.54	76	8.71	86	8.88	96	9.06	9.0
10.0	55	9.39	65	9.54	76	9.70	86	9.88	96	10.01	10.0

Figure 8–5 *Beaufort scale 0: under 1 knot.*

Figure 8–6 *Beaufort scale 1: 1–3 knots.*

Figure 8–7 *Beaufort scale 2: 4–6 knots.*

Figure 8–8 *Beaufort scale 3: 7–10 knots.*

Figure 8–9 *Beaufort scale 4: 11–16 knots.*

Figure 8–10 *Beaufort scale 5: 17–21 knots.*

Figure 8–11 *Beaufort scale 6: 22–27 knots.*

Figure 8–12 *Beaufort scale 7: 28–33 knots.*

Figure 8–13 *Beaufort scale 8: 34–40 knots.*

Figure 8–14 *Beaufort scale 9: 41–47 knots.*

Figure 8–15 *Beaufort scale 10: 48–55 knots.*

Figure 8–16 *Beaufort scale 11: 56–63 knots.*

Figure 8–17 *Beaufort scale 12: 64–71 knots.*

In the United States it was formerly called the **centigrade** scale.

Réaumur temperature is based upon a scale in which water freezes at 0° and boils at 80°.

Absolute zero is considered to be the lowest possible temperature, at which there is no molecular motion and a body has no heat. For some purposes, it is convenient to express temperature by a scale at which 0° is absolute zero. This is called **absolute** temperature. If Fahrenheit degrees are used, it may be called **Rankine (R)** temperature; and if Celsius, **Kelvin (K)** temperature. The Kelvin scale is more widely used than the Rankine. Absolute zero is at $(-)459°.67$ F or $(-)273°.15$ C.

Temperature by one scale can be converted to that at another by means of the relationship that exists between the scales. Thus,

$$C = \frac{5}{9}(F - 32)$$

and

$$F = \frac{9}{5}C + 32.$$

Temperature measurement is made by means of a **thermometer.** Most thermometers are based upon the principle that materials expand with increase of temperature, and contract as temperature decreases. In its most usual form a thermometer consists of a bulb filled with mercury and connected to a tube of very small cross-sectional area. The mercury only partly fills the tube. In the remainder is a vacuum created during construction of the instrument. The air is driven out by boiling the mercury, and the top of the tube is then sealed by a flame. As the mercury expands or contracts with changing temperature, the length of the mercury column in the tube changes. Temperature is indicated by the position of the top of the column of mercury with respect to a scale etched on the glass tube or placed on the thermometer support.

A maximum thermometer has a constriction in the tube, near the bulb. As temperature increases, the expanding mercury is forced past the constriction, but will not return as temperature decreases. Thus, it indicates the highest temperature which has occurred since the last setting. This principle is utilized in clinical thermometers, used for measuring body temperature. The mercury can be forced back into the bulb by centrifugal force applied by swinging the arm rapidly. Meteorologists have a device called a "Townsend support" for accomplishing this with less leffort and less possibility of breakage.

A minimum thermometer uses alcohol instead of mercury. The upper part of the tube contains air under slight pressure, to prevent evaporation of the alcohol with resultant "breaks" in the column as the alcohol later condenses. The thermometer contains an index which is so constructed as to allow alcohol to flow past it up the tube with rising temperatures, but which moves downward in the tube if the temperature falls below it, being drawn down by the effect of surface tension exerted by the bottom of the meniscus (curved upper surface) of the column of alcohol as it reaches the index. Due to this effect, the index remains at the lowest temperature which has occurred since the last setting. Setting is accomplished by tilting the thermometer until the bulb is uppermost, when the index returns to the current temperature. The thermometer is normally maintained at an angle of about $5°$ to the horizontal, with the bulb at the lower end. A Townsend support is used for this purpose.

Temperature can be measured by means of a **thermograph** (figure 8–19) which is a recording thermometer. In its outward appearance this instrument is similar to a barograph. The pen arm is connected, through a linkage, to the thermometric element, which usually consists of a metal tube shaped in the form of an arc and containing alcohol. As the alcohol expands with temperature increase, it tends to straighten the tube; and as the temperature decreases, the con-

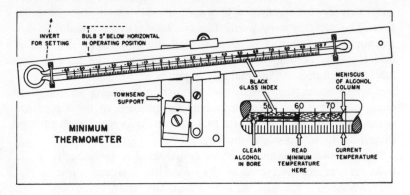

Figure 8–18 *A minimum thermometer.*

tracting alcohol permits the tube to resume its curved shape. The linkage magnifies these variations and transmits them to the pen, which records the temperature on a chart placed around a clock-driven, revolving cylinder.

The freezing point of mercury is about $(-)38°$ F. Various substances are used to measure lower temperatures, the most common being some form of alcohol, which has a freezing point well below

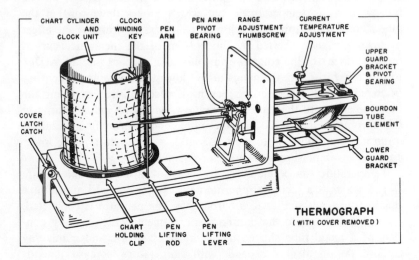

Figure 8–19 *A thermograph with cover removed.*

$(-)100°$ F. For even lower temperatures, below those ever recorded in the atmosphere, gas may be used instead of a liquid. Thermometers based upon other principles, such as unequal expansion of dissimilar metals, melting point of a substance, color, etc., are sometimes used, particularly for temperatures considerably higher or much lower than those occurring in the atmosphere.

Temperature measuring equipment should be placed in a shelter which protects it from mechanical damage and direct rays of the sun. The shelter should have louvered sides to permit free access of air. Aboard ship, the shelter should be placed in an exposed position as far as practicable from metal bulkheads. On vessels where shelters are not available, the temperature measurement should be made in shade at an exposed position on the windward side.

If the temperature of sea water at the surface is desired, a sample should be obtained by bucket, preferably a canvas bucket, from a forward position well clear of any discharge lines. The sample should be taken immediately to a place where it is sheltered from wind and sun. The water should then be stirred with the thermometer, keeping the bulb submerged, until an essentially constant reading is obtained.

Humidity is the condition of the atmosphere with reference to its water vapor content. **Absolute humidity** is a measure of the mass of vapor per unit volume of air. **Relative humidity** is the ratio (stated as a percentage) of the existing vapor pressure to the vapor pressure corresponding to saturation at the prevailing temperature and atmospheric pressure. This is very nearly the ratio of the amount of water vapor present to the amount that the air could hold at the same temperature and pressure if it were saturated.

As air cools, its capacity for holding water vapor decreases. Therefore, as air temperature decreases, the relative humidity increases. At some point, saturation takes place, and any further cooling results in condensation of some of the moisture. The temperature at which this occurs is called the **dew point,** and the moisture deposited upon natural objects is called **dew** if it forms in the liquid state, or **frost** if it forms in the frozen state.

The same process causes moisture to form on the outside of a container of cold liquid, the liquid cooling the air in the immediate vicinity of the container until it reaches the dew point. When moisture is deposited on man-made objects, it is usually called **sweat.** It occurs whenever the temperature of a surface is lower than the dew point of the air in contact with it. It is of particular concern to the mariner because of its effect upon his instruments, and possible

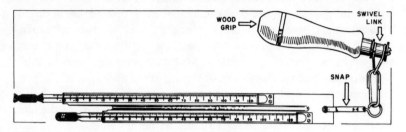

Figure 8–20 *A sling psychrometer.*

damage to his ship. Lenses of optical instruments may sweat, usually with such small droplets that the surface has a "frosted" appearance. When this occurs, the instrument is said to "fog" or "fog up," and is useless until the moisture is removed. Damage is often caused by corrosion or direct water damage when pipes sweat and drip, or when the inside of the shell plates of a vessel sweat.

Clouds and fog form by "sweating" of minute particles of dust, salt, etc., in the air. Each particle forms a nucleus around which a droplet of water forms. If air is completely free from solid particles on which water vapor may condense, the extra moisture remains in the vapor state, and the air is said to be **supersaturated.**

Relative humidity and dew point are measured by means of a **hygrometer.** The most common type, called a **psychrometer** consists of two thermometers mounted together on a single strip of material, as shown in figure 8–20. One of the thermometers is mounted a little lower than the other, and has its bulb covered with muslin. When the muslin covering is thoroughly moistened and the thermometer well ventilated, evaporation cools the bulb of the thermometer, causing it to indicate a lower reading than the other. A **sling psychrometer,** illustrated in figure 8–20, is ventilated by whirling the thermometers. Some psychrometers use a fan. **Dry-bulb temperature** is indicated by the uncovered **dry-bulb thermometer,** and **wet-bulb temperature** is indicated by the muslin-covered **wet-bulb thermometer.** The difference between these two temperatures, and the dry-bulb temperature, are used to find the relative humidity and dew point.

A hygrothermograph combines the features of both the hygrograph and the thermograph, providing a continuous record of both relative humidity and air temperature for seven days on a single chart. It has the same limitations as the hygrograph and the thermograph and its indications should be checked daily by psychrometer and thermometer.

Visibility measurement—Visibility is the extreme horizontal distance at which prominent objects can be seen and identified by the unaided eye. It is usually measured directly by the human eye. Ashore, the distances of various buildings, trees, lights, and other objects are measured and used as a guide in estimating the visibility. At sea, however, such an estimate is difficult to make with accuracy. Other ships and the horizon may be of some assistance.

Upper Air Observations—Upper air information provides the third dimension to the weather map. The equipment necessary to obtain such information is quite expensive, and the observations are time consuming. Consequently, the network of observing stations is quite sparse compared to that for surface observations, particularly over the oceans and in isolated land areas. Where facilities exist, upper air observations are made by means of unmanned balloons in conjunction with theodolites, radiosondes, radar, and radio direction finders. Observations are sometimes made by aircraft.

Pilot balloons are free balloons released at the surface of the earth and followed by optical means to determine their movement in relation to the point from which released. They are of neoprene latex (occasionally of natural rubber latex) a few thousandths of an inch thick, and have a nominal weight of either 30 or 100 grams. The balloons are inflated with helium or hydrogen to a definite free-lift capacity for which ascensional rate tables have been prepared. The neck of each balloon is then securely fastened to prevent leakage of the gas, and the balloon is released. A theodolite is trained on the balloon, which is kept in the field of vision of the instrument throughout the observation.

By means of a buzzer signal the observer is warned five seconds prior to the end of each minute after release. The cross hairs of the theodolite are then brought to bear on the balloon at the end of each minute (also signalled by the buzzer), and the horizontal and vertical angles are read to the nearest tenth of a degree. These data are then plotted on polar coordinate paper similar to a maneuvering board, and the wind speed and direction at each selected level (each 1,000-foot level) are determined.

An observation of winds aloft made in this manner is called a **pibal,** from **pi**lot **bal**loon observation. If the same procedure is used with a sounding balloon, the observation is called a **rabal,** from **ra**dio **bal**loon observation.

The shipboard-type theodolite (figure 8–21) is mounted on gimbals atop a tripod. A counterbalance is provided to serve as a pendulum in maintaining the instrument approximately horizontal. The instrument

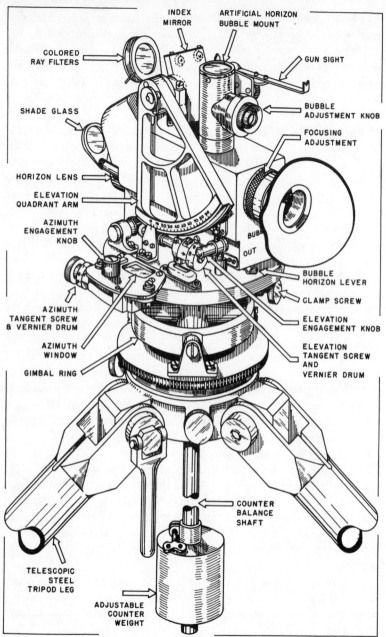

Figure 8–21 *A shipboard-type meteorological theodolite.*

is aligned with the longitudinal axis of the craft, so that *relative* bearings are observed. Elevation angles are measured in a manner similar to the measurement of altitudes of celestial bodies, an image of the balloon being brought into coincidence with the direct view of the horizon. A bubble artificial horizon is also provided.

Radiosondes are miniature radio transmitters carried aloft by **sounding balloons** which ascend at the rate of about 1,000 feet per minute, to a height of nearly 100,000 feet. The transmitter, powered by a compact battery, transmits on a frequency of 72, 403, or 1,680 megacycles per second. In the United States the 72-megacycle instruments have been replaced by 403-megacycle radiosondes.

As the radiosonde ascends, it transmits a continuous-wave radio signal on its assigned frequency. This signal is modulated by pressure, temperature, and relative humidity in turn.

The transmitted radio signals are received by an antenna and radio receiver at the surface. They are fed through an electronic frequency meter, and then recorded. By this means a continuous record is made to the height at which the balloon bursts or its signals can no longer be received.

An observation made in this way is called a **raob,** from **ra**diosonde **ob**servation.

Electronic Measurement of Winds Aloft—If either a pilot balloon or sounding balloon is fitted with a metal target and tracked by radar, height, slant distance, and bearing are available, permitting determination of wind speed and direction. Radio direction finder equipment which permits measurement of both horizontal and vertical directions has been developed and is in use ashore for tracking radiosondes. Similar equipment for use aboard ship is under development. An observation made by tracking with either radar or radio direction finder is called a **rawin,** from **ra**dio **win**ds-aloft observation. A combined raob and rawin is called a **rawinsonde.**

Observations by Aircraft—Reports from aircraft are helpful in making upper air observations. By this means, winds, heights of clouds, visibility, etc., can be determined. An aircraft flying over the ocean and equipped with both absolute and barometric altimeters can supply valuable information on the height of the pressure level at which it is flying. Such reports are used in connection with pressure pattern navigation. They are also useful in establishing positions of high and low pressure centers.

The Air Weather Service of the U. S. Department of Defense makes regular flights to collect weather information. These flights are made along established routes over the oceans and in the arctic where adequate coverage is not otherwise available. In addition, the U. S. Navy and U. S. Air Force, in cooperation with the Weather Bureau, make flights into tropical cyclones.

Precipitation Measurement—Any type of condensed water vapor that falls to the earth's surface is called **precipitation.** It may be liquid, freezing, or frozen when it arrives at the surface. Measurement of precipitation normally includes only the determination of the amount of rain or snow that has fallen in a given period of time. For purposes of comparison, snow measurement is obtained by melting the snow to its water equivalent. Depth of snow is also measured to determine the amount of snowfall.

The usual type of **nonrecording precipitation gage** consists of a collector ring, funnel, and measuring cylinder set within a receiver. All precipitation falling on the area encompassed by the collector ring descends through the funnel into the measuring cylinder, where it is measured directly by means of a rod graduated in tenths of an inch. Since the cross-sectional area of the measuring cylinder is exactly one-tenth that of the collector ring, each 0.1 inch collected is a measure of 0.01 inch of precipitation. When precipitation is in the form of snow, the measuring tube is removed, permitting the snow to collect in the larger receiver. The receiver is placed in a container of warm water until the snow melts. The resulting liquid is then poured into the measuring tube and measured.

The most representative measurement of precipitation from snow is obtained by removing the collector ring and funnel, and using a slat screen to reduce the effect of wind.

Automatic weather stations provide regularly scheduled transmissions of meteorological measurements by radio. They are used at isolated and relatively inaccessible locations from which weather data are of great importance to the weather forecaster. The measurements usually obtained are of wind speed and direction, atmospheric pressure, temperature, and relative humidity.

Recording Observations—Aboard ship, weather observations are recorded on the *Ship Weather Observation Sheet* (figure 8–22).

The symbols used in the "weather" column are as follows:

CLR—Clear or a few clouds
SCT—Scattered clouds—0.1 to 0.5 clouds
BKN—Broken clouds—0.6 to 0.9 clouds
OVC—Overcast—more than 0.9 clouds

OPNAV FORM 3144-1 (9-64)

DEPARTMENT OF THE NAVY

SHIP WEATHER OBSERVATION SHEET

USS ______________________ DATE (GMT) ______________ 19 ______

AT/PASSAGE FROM ______________________ TO ______________________

TABLE I

TIME (GMT)	WINDS ☐ ✓ IF ESTIMATED		VISI-BIL-ITY (Miles)	WEATHER (Symbols)	BAROMETER (Inches)	TEMPERATURE (Degrees and tenths)		CLOUDS			SEA WATER TEMP. (Degrees and tenths)	SEA WAVES			SWELL WAVES		
	Direction (True)	Force (Knots)				Dry Bulb	Wet Bulb	Amount (Tenths)	Height	Type		Direction (True)	Period (Seconds)	Height (Feet)	Direction (True)	Period (Seconds)	Height (Feet)
00																	
01																	
02																	
03																	
04																	
05																	
06																	
07																	
08																	
09																	
10																	
11																	
12																	
13																	
14																	
15																	
16																	
17																	
18																	
19																	
20																	
21																	
22																	
23																	

TABLE II
SYNOPTIC OBSERVATIONS

FIRST GROUP OF MESSAGE	Day of Week (1-7) (GMT)	POSITION OF SHIP			TIME (GMT)	Total Cloud Amt. (Coded)	WIND		Visi-bil-ity (90-99)	WEATHER		PRESSURE	AIR TEMP. (°C)	CLOUDS					Course of Ship (0-9)	Speed of Ship (0-9)	3-HOUR PRESSURE TENDENCY		SIGNIFICANT CLOUD			
		Octant (0-3) (5-8)	Latitude (Degrees and tenths)	Longitude (Degrees and tenths)			Direction (True) (00-36)	Speed (True) (Knots)		Present (00-99)	Past (0-9)	Barometer Corrected (Mb)		Amount of Low Cloud	Type of C_L (0-9)	Height of Low Cloud	Type of C_M (0-9)	Type of C_H (0-9)			Characteristic (0-8)	Amount of Change (Mb and tenths)	Indicator	Amount (Eights)	Type	Height
1	2	3	4	5	6	7	8	9	10	11	12	13	14	15	16	17	18	19	20	21	22	23	24	25	26	27
	Y	Q	$L_a L_a L_a$	$L_o L_o L_o$	GG	N	dd	ff	VV	ww	W	ppp	TT	N_h	C_L	h	C_M	C_H	D_s	V_s	a	pp	8	N_s	C	$h_s h_s$
SHIP					00																		8			
SHIP					06																		8			
SHIP					12																		8			
SHIP					18																		8			

Indicator	AIR-SEA DIFF. (Coded)	DEW POINT (°C)	SEA WAVES				SWELL WAVES				ICE ACCRETION				SEA ICE					
			Indicator	Direction (Coded)	Period (Coded)	Height (Coded)	Indicator	Direction (Coded)	Period (Coded)	Height (Coded)	Indicator	Source	Thickness	Rate	Indicator	Kind	Effect	Bearing	Distance	Orientation
28	29	30	31	32	33	34	35	36	37	38	39	40	41	42	43	44	45	46	47	48
0	$T_s T_s$	$T_d T_d$	1	$d_w d_w$	P_w	H_w	1	$d_w d_w$	P_w	H_w	2	I_s	$E_s E_s$	R_s	ICE	C_2	K	D_i	r	e
0			1				1				2				ICE					
0			1				1				2				ICE					
0			1				1				2				ICE					
0			1				1				2				ICE					

DO NOT TRANSMIT		
Dry Bulb (Degrees and tenths)	Wet Bulb (Degrees and tenths)	Sea Water Temp. (Degrees and tenths)
A_1	A_2	A_3
Celsius	Celsius	Celsius

REMARKS ______________________ EXAMINED ______________________ USN, NAVIGATOR

Figure 8–22 *Ship Weather Observation Sheet.*

T—Thunderstorm
R—Rain
RW—Rain showers
L—Drizzle
ZR—Freezing rain
ZL—Freezing drizzle
E—Sleet
F—Fog
GF—Shallow fog (ground fog)
EW—Sleet showers
S—Snow
SW—Snow showers
IC—Ice crystals
A—Hail
IF—Ice fog
H—Haze
K—Smoke
D—Dust
BY—Blowing spray

=9=

Clouds

Clouds are visible assemblages of numerous tiny droplets of water, or ice crystals, formed by condensation of water vapor in the air, with the bases of the assemblages above the surface of the earth. **Fog** is a similar assemblage in contact with the surface of the earth.

The shape, size, height, thickness, and nature of a cloud depend upon the conditions under which it is formed. Therefore, clouds are indicators of various processes occurring in the atmosphere. The ability to recognize different types and a knowledge of the conditions associated with them are useful in predicting future weather. (See Cloud Formation Insert.)

Although the variety of clouds is virtually endless, they may be classified according to general type. Clouds are grouped generally into four "families" according to some common characteristic. **High clouds** are those having a mean lower level above 20,000 feet. They are composed principally of ice crystals. **Middle clouds** have a mean level between 6,500 and 20,000 feet. They are composed largely of water droplets, although the higher ones have a tendency toward ice particles. **Low clouds** have a mean upper level of less than 6,500 feet. These clouds are composed entirely of water droplets.

Clouds with vertical development are a distinctive group formed by rising air which is cooled as it reaches greater heights. When it reaches the height of the dew point, some of its water vapor condenses. Therefore, the bottoms of such clouds are usually flat. Clouds with vertical development may begin at almost any level, but generally within the low cloud range. They may extend to great heights, well above the lower limit of high clouds. They form as water droplets, but toward the top they may freeze.

Within these four families are ten principal cloud types. The names of these are composed of various combinations and forms of the following basic words, all from Latin:

Cirrus, meaning "curl."
Cumulus, meaning "heap."
Stratus, meaning "layer."
Alto, meaning "high."
Nimbus, meaning "rain."

The first three are the basic cloud types. Individual cloud types recognize certain characteristics, variations, or combinations of these. The ten principal cloud types are:

High clouds. Cirrus (Ci) are detached high clouds of delicate and fibrous appearance, without shading, generally white in color, and often a silky appearance. Their fibrous and feathery appearance is due to the fact that they are composed entirely of ice crystals. Cirrus appear in varied forms such as isolated tufts; long, thin lines across the sky; branching, feather-like plumes; curved wisps which may end in tufts, etc. These clouds may be arranged in parallel bands which cross the sky in great circles and appear to converge toward a point on the horizon. This may indicate, in a general way, the direction of a low pressure area. Cirrus may be brilliantly colored at sunrise and sunset. Because of their height, they become illuminated before other clouds in the morning, and remain lighted after others at sunset. Cirrus are generally associated with fair weather, but if they are followed by lower and thicker clouds, they are often the forerunner of rain or snow.

Cirrocumulus (Cc) are high clouds composed of small white flakes or scales, or of very small globular masses, usually without shadows and arranged in groups or lines, or more often in ripples resembling those of sand on the seashore. One form of cirrocumulus is popularly known as "mackerel sky" because the pattern resembles the scales on the back of a mackerel. Like cirrus, cirrocumulus are composed of ice crystals and are generally associated with fair weather, but may precede a storm if they thicken and lower. They may turn gray and appear hard before thickening.

Cirrostratus (Cs) are thin, whitish, high clouds sometimes covering the sky completely and giving it a milky appearance and at other times presenting, more or less distinctly, a formation like a tangled web. The thin veil is not sufficiently dense to blur the outline of sun or moon. However, the ice crystals of which the cloud is composed refract the light passing through in such a way that halos may form with the sun or moon at the center. Cirrus thickening and changing into cirrostratus is popularly known as "mares' tails." If it continues to thicken and lower, the ice crystals melting to form water droplets,

International Cloud Code Guide For The Mariner

Family "A," High Clouds: Cirrus (Ci), Cirrocumulus (Cc), Cirrostratus (Cs). Mean lower level, 6000 meters, 20,000 feet.

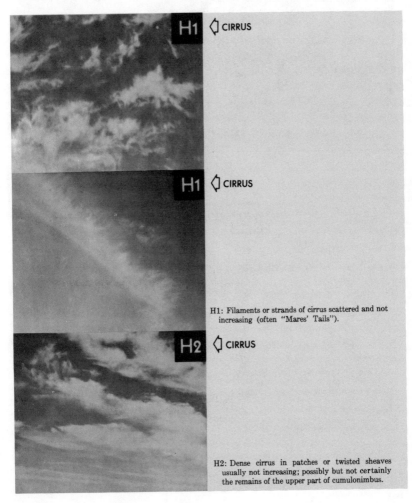

H1 ◁ CIRRUS

H1 ◁ CIRRUS

H1: Filaments or strands of cirrus scattered and not increasing (often "Mares' Tails").

H2 ◁ CIRRUS

H2: Dense cirrus in patches or twisted sheaves usually not increasing; possibly but not certainly the remains of the upper part of cumulonimbus.

H3 ◁ CIRRUS

H3: Cirrus, often anvil-shaped; either the remains of the upper portions of cumulonimbus or part of a distant cumulonimbus, the rest of which is not visible. (If there is doubt as to the cumulonimbus origin or association, code H2 should be used.)

(See notes on L9 for coding requirements when cumulonimbus is present.)

H4 ◁ CIRRUS

H4: Cirrus (often hook-shaped) gradually spreading over the sky and usually thickening as a whole.

The essential characteristic is the gradual spreading over the sky. Even if the clouds are hook-shaped but are not spreading over the sky, they should be coded as H1.

Note that these clouds must extend to the horizon from which they are advancing, where owing to the effect of perspective they may assume the appearance of cirrostratus.

CIRRUS & CIRROSTRATUS ▷ **H5**

H5: Cirrus and cirrostratus, often in bands converging toward the horizon; or cirrostratus alone; in either case gradually spreading over the sky and usually thickening as a whole, but the continuous layer not reaching 45° altitude.

When cirrus is present, the angular altitude refers to the leading edge of the cirrostratus layer.

CIRRUS & CIRROSTRATUS ▷ **H6**

H6: Cirrus and cirrostratus often in bands converging toward the horizon; or cirrostratus alone; in either case gradually spreading over the sky and usually thickening as a whole, and the continuous layer exceeding 45° altitude.

When cirrus is present, the angular altitude refers to the leading edge of the cirrostratus layer.

CIRROSTRATUS ▷ H7

H7: Cirrostratus covering the entire sky.

During the day, when the sun is sufficiently high above the horizon, the sheet is never thick enough to prevent shadows of objects on the ground.

If lower clouds are present and cirrostratus is visible, the cirrostratus will be coded H7 or H8 according as, in the opinion of the observer, the sky is or is not completely covered with cirrostratus.

CIRROSTRATUS ▷ H8

H8: Cirrostratus not increasing and not covering the whole sky; cirrus and cirrocumulus may be present.

If cirrocumulus is present, the cirrostratus must predominate to satisfy the requirements of H8. If the cirrocumulus predominates, the sky would be coded as H9.

CIRROCUMULUS ▷ H9

H9: Cirrocumulus alone or cirrocumulus with some cirrus or cirrostratus, but the cirrocumulus being the main cirriform cloud present. (Cirrocumulus may be present in H1 to H8.)

Family "B," Middle Clouds: Altocumulus (Ac), Altostratus (As). Mean upper level, 6000 meters, 20,000 feet; mean lower level, 2000 meters, 6500 feet.

M1 ◁ ALTOSTRATUS

M1: Thin altostratus (semi-transparent everywhere) through which the sun or moon can be dimly seen. A sheet of this cloud resembles thick cirrostratus from which it is often derived without any break; but halo phenomena, sun pillar, etc., are not seen in cirrostratus, and the sun or moon appears as though shining through ground glass and does not cast shadows.

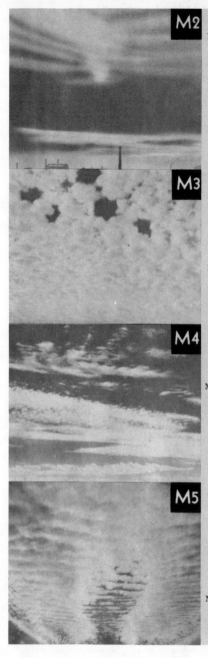

M2 ◁ ALTOSTRATUS OR NIMBOSTRATUS

M2: Thick altostratus or nimbostratus (through portions of the sheet the position of the sun or moon may be indicated by a light patch).

The sun and moon are completely hidden by at least some parts of the cloud sheet, which may be fibrous in appearance. Thick altostratus can be formed either by thickening of thin altostratus or by the fusing together of cloudlets in a sheet of altocumulus.

Nimbostratus is derived either by a change from thick altostratus or by the fusing together of the cloud elements in a sheet of dense alto-cumulus, stratocumulus, or stratus.

When nimbostratus gives precipitation it is in
(Continued)

M3 ◁ ALTOCUMULUS

the form of continuous rain or snow. Nim-bostratus usually has a dark gray color and its lower surface always has a wet appearance, widespread trailing precipitation, "virga," which may or may not reach the ground; it is quite uniform and it is not possible to make out definite detail.

M3: Thin (semi-transparent) altocumulus; cloud elements not changing much; at a single level.

This cloud is fairly regular and of uniform thick-ness. The cloudlets or waves are always separated by clear spaces or lighter patches and are neither very large nor very dark.

M4 ◁ ALTOCUMULUS

M4: Thin (semi-transparent) altocumulus in patches (often almond or fish-shaped); cloud elements continually changing and/or occurring at more than one level.

Lenticular patches properly coded as M4 often pile up in layers, at times with clear spaces between. When they merge horizontally in the form of rafts or somewhat discontinuous sheets, they must be coded M3 if they are not gradually spreading over the sky, and M5 if they are.

M5 ◁ ALTOCUMULUS

M5: Thin (semi-transparent) altocumulus in bands or in a layer gradually spreading over the sky and usually thickening as a whole; it may become partly opaque or double-layered. *(Continued)*

ALTOCUMULUS ▷

M5 designates one or perhaps two advancing layers of altocumulus, usually of irregular thickness, the amount and thickness of which are definitely increasing. The altocumulus stretches to the horizon, at least in the direction from which it is advancing.

M6: Altocumulus formed by the spreading out of cumulus.

Cumulus clouds of sufficiently great vertical development may undergo an extension of their summits while their bases may gradually melt away. Sheets of altocumulus, which are generally fairly thick and opaque at first, are formed in
(Continued)

ALTOCUMULUS ▷

this manner. They have rather large elements, dark and soft; later they may thin out and finally have rifts in them or, at any rate, semi-transparent intervening spaces.

When there is doubt as to whether a spreading sheet should be termed alto-culumus or stratocumulus, it is best to code M6, since the cumulus clouds may then also be coded (L1 or L2).

M7: Any of the following cases:
    (a) Double-layered altocumulus, usually opaque in parts, not increasing.
    (b) A thick (opaque) layer of altocumulus, not increasing.
(Continued)

ALTOSTRATUS & ALTOCUMULUS ▷

    (c) Altostratus and altocumulus both present at the same or different levels.

Type *(a)*: Two layers of altocumulus, the lower of which resembles a gray veil, often hardly visible, lying at a level very little lower in places and for a short time hiding the cloudlets of the altocumulus sheet sufficiently to give it the appearance of altostratus. This double-layered altocumulus is coded M7 only if the altocumulus is not systematically increasing; otherwise it is coded M5.

Type *(b)*: The under surface of opaque altocumulus is marked by a more or less corrugated or wave-like structure, sometimes called wrinkled.

ALTOCUMULUS ▷

M8: Altocumulus in the form of cumulus-shaped tufts or altocumulus with turrets.

Tufted altocumulus are white or gray cloudlets that have no definite shadows and that have very slightly domed tops. The tufts resemble very small broken cumulus clouds whose bases are not flat. Turreted altocumulus shows somewhat greater vertical development than the tufted form. The turrets rise from a common flat base.

If either turreted or tufted altocumulus is present, even in small amounts, with another type of altocumulus or with altostratus predominating, M8 must be coded, except when the sky is chaotic and M9 is therefore required.

ALTOCUMULUS ▷ 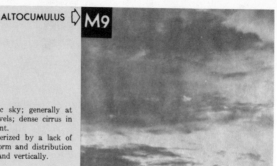 M9

M9: Altocumulus of a chaotic sky; generally at more than two different levels; dense cirrus in patches is usually also present.

These clouds are characterized by a lack of regularity with respect to form and distribution in space, both horizontally and vertically.

Family "C," Low Clouds: Stratocumulus (Sc), Stratus (St), Nimbostratus (Ns). Mean upper level, 2000 meters, 6500 feet; mean lower level, close to surface.

 L4 ◁ STRATOCUMULUS

L4: Stratocumulus formed by the spreading out of cumulus; cumulus also often present.

STRATOCUMULUS ▷ L5

L5: Stratocumulus not formed by the spreading out of cumulus.

This includes a wide variety of aspects of stratocumulus, ranging from thin clouds at a single level with semi-transparent parts or even clear spaces, to the dark and menacing clouds, often at two or more levels, immediately before or after precipitation. Code L5 applies only in the absence of fractocumulus of bad weather, cumulus, or cumulonimbus. If stratus and stratocumulus are both present, code L5 applies while stratocumulus is dominant.

STRATUS OR FRACTOSTRATUS ▷ L6

L6: Stratus or Fractostratus, or both, but not fractostratus of bad weather.

These clouds are usually in a low single layer and may be localized in extent. Clouds properly coded L6, unlike those coded L7, are not very dark or menacing.

The designation "fractostratus" is used when a layer of stratus is broken up into irregular shreds.

L7: Fractostratus and/or fractocumulus of bad weather ("scud") usually under altostratus and nimbostratus. (By "bad weather" is meant the conditions usually prevailing immediately before, during, or immediately after precipitation.)

Fractocumulus clouds of bad weather are usually dark, receive little light; these clouds generally become very numerous and may merge into a sheet covering the entire sky.

(See L1 for description of fractocumulus of good weather.)

Family "D," Clouds with Vertical Development: Cumulus (Cu), Cumulonimbus (Cb).

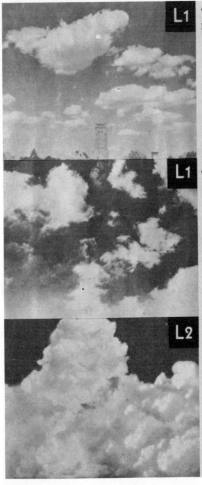

L1 ◁ CUMULUS

L1: Cumulus with little vertical development and seemingly flattened.

These clouds occur in three forms:
   A. In a state of formation.
   B. Completely formed.
   C. Completely formed but broken up by the wind (fractocumulus).

They usually have a marked diurnal growth over land, developing until the middle of the afternoon and decreasing later, both as to amount and vertical extent. At sea and on coasts, cumulus clouds often occur at night.

Note that even though cumulus clouds with little vertical development should dominate the
(Continued)

L1 ◁ CUMULUS

sky, the presence of any cumulus of considerable vertical development will require coding L2. The presence of even a single cumulonimbus with any amount of stratocumulus, stratus or cumulus clouds will require coding L3 or L9:

When the cumulus clouds begin to spread out in any part of the sky, the clouds will be coded L4 rather than L1, unless the spreading portions form altocumulus, in which case they will be coded M6 and L1.

Fractocumulus of fine weather (L1) are detached white clouds usually in an otherwise clear sky. (See L7 for description of fracto-cumulus of bad weather.)

L2 ◁ CUMULUS

L2: Cumulus of considerable development, generally towering, with or without other cumulus or stratocumulus; bases all at the same level.

These clouds are massive in appearance, occasionally wind-tossed and broken, with horizontal bases and very great vertical development. They are sometimes in the form of towers or of complex heaps with "cauliflower" formation. They often have caps or hoods (pileus), which are distinguished from the spreading tops of cumu-lonimbus by their smoothness, sharpness, and short duration (a few minutes).

(See L8 for the coding of cumulus of consider-able development and stratocumulus with bases at different levels.)

L3 ◁ CUMULONIMBUS

L3: Cumulonimbus with tops lacking clear-cut outlines but distinctly not cirriform or anvil-shaped; with or without cumulus, stratocumulus, or stratus.

These are cumuliform clouds of great vertical development with tops composed in part at least of ice crystals, the presence of which is revealed by a partial or general indefiniteness of previously well-defined "cauliflower" tops.

(See L9 for the coding of cumulonimbus having clearly fibrous tops.)

CUMULUS & STRATOCUMULUS ▷ L8

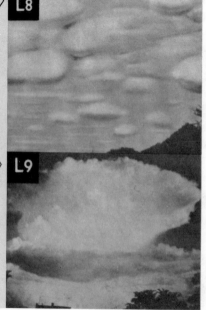

L8: Cumulus and stratocumulus other than those formed by the spreading out of cumulus with bases at different levels. The lower cumulus clouds may or may not extend up through the upper stratocumulus layer.

L9: Cumulonimbus having a clearly fibrous (cirriform) top, often anvil-shaped, with or without cumulus, stratocumulus, or stratus, or "scud."

By extension at various levels, cumulonimbus
(Continued)

CUMULONIMBUS ▷ L9

often produces cirrus, altocumulus, or stratocumulus clouds. Therefore, cumulonimbus may coexist with clouds that should be coded, when detached from the parent cloud, H2, H3, or M6. If these clouds are not detached, they should not be coded separately.

Cumulonimbus clouds generally produce showers of rain or snow and sometimes of hail, and often thunderstorms as well.

If the whole of the cloud cannot be seen the fall of a real heavy shower is enough to characterize the cloud as a cumulonimbus.

the cloud formation is known as altostratus. When this occurs, rain may normally be expected within 24 hours. The more brushlike the cirrus when the sky appears, the stronger the wind at the level of the cloud.

Middle clouds. Altocumulus (Ac) are middle clouds consisting of a layer of large, ball-like masses that tend to merge together. The balls or patches may vary in thickness and color from dazzling white to dark gray, but they are more or less regularly arranged. They may appear as distinct patches similar to cirrocumulus but can be distinguished by the fact that individual patches are generally larger, and show distinct shadows in some places. They are often mistaken for stratocumulus. If this form thickens and lowers, it may produce thundery weather and showers, but it does not bring prolonged bad weather. Sometimes the patches merge to form a series of big rolls that resemble ocean waves, but with streaks of blue sky. Because of perspective, the rolls appear to run together near the horizon. These regular parallel bands differ from cirrocumulus in that they occur in larger masses with shadows. These clouds move in the direction of the short dimension of the rolls, as do ocean waves. Sometimes altocumulus appear briefly before a thunderstorm. They are generally arranged in a line with a flat horizontal base, giving the impression of turrets on a castle. The turreted tops may look like miniature cumulus and possess considerable depth and great length. These clouds usually indicate a change to chaotic, thundery skies.

Altostratus (As) are middle clouds having the appearance of a grayish or bluish, fibrous veil or sheet. The sun or moon, when seen through these clouds, appears as if it were shining through ground glass, with a corona around it. Halos are not formed. If these clouds thicken and lower, or if low, ragged "scud" or rain clouds (nimbostratus) form below them, continuous rain or snow may be expected within a few hours.

Low clouds. Stratocumulus (Sc) are low clouds composed of soft, gray, roll-shaped masses. They may be shaped in long, parallel rolls similar to altocumulus, moving forward with the wind. The motion is in the direction of their short dimension, like ocean waves. These clouds, which vary greatly in altitude, are the final product of the characteristic daily change that takes place in cumulus clouds. They are usually followed by clear skies during the night.

Stratus (St) is a low cloud in a uniform layer resembling fog. Often the base is not more than 1,000 feet high. A veil of thin stratus gives the sky a hazy appearance. Stratus is often quite thick, permitting so little sunlight to penetrate that it appears dark to an observer below it. From above, it looks white. Light mist may descend from stratus. Strong wind sometimes breaks stratus into shreds called "fracto-stratus."

Nimbostratus (Ns) is a low, dark, shapeless cloud layer, usually nearly uniform, but sometimes with ragged, wet-looking bases. Nimbostratus is the typical rain cloud. The precipitation which falls from this cloud is steady or intermittent, but not showery.

Clouds with vertical development. Cumulus (Cu) are dense clouds with vertical development. They have a horizontal base and dome-shaped upper surface, with protuberances extending above the dome. Cumulus appear in small patches, and never cover the entire sky. When the vertical development is not great, the clouds appear in patches resembling tufts of cotton or wool, being popularly called "woolpack" clouds. The horizontal bases of such clouds may not be noticeable. These are called "fair weather" cumulus because they always accompany good weather. However, they may merge with alto-cumulus, or may grow to cumulonimbus before a thunderstorm. Since cumulus are formed by updrafts, they are accompanied by turbulence, causing "bumpiness" in the air. The extent of turbulence is proportional to the vertical extent of the clouds. Cumulus are marked by strong contrasts of light and dark.

Cumulonimbus (Cb) is a massive cloud with great vertical development, rising in mountainous towers to great heights. The upper part consists of ice crystals, and often spreads out in the shape of an anvil which may be seen at such distances that the base may be below the horizon. Cumulonimbus often produces showers of rain, snow, or hail, frequently accompanied by thunder. Because of this, the cloud is often popularly called a "thundercloud" or "thunderhead." The base is horizontal, but as showers occur it lowers and becomes ragged.

Cloud Height Measurement—At sea, cloud heights are often determined by estimate. This is a difficult task, particularly at night. A searchlight may be of some assistance.

—10—

Tropical Cyclones

A tropical cyclone is a violent cyclone originating in the tropics. Although it generally resembles the extratropical cyclone originating in higher latitudes, there are important differences, the principal one being the concentration of a large amount of energy into a relatively small area. Tropical cyclones are infrequent in comparison with middle- and high-latitude storms, but they have a record of destruction far exceeding that of any other type of storm. Because of their fury, and the fact that they are predominantly oceanic, they merit the special attention of all mariners, whether professional or amateur.

Rarely does the mariner who has experienced a fully developed tropical cyclone at sea wish to encounter a second one. He has learned the wisdom of avoiding them if possible. The uninitiated may be misled by the deceptively small size of a tropical cyclone as it appears on a weather map, and by the fine weather experienced only a few hundred miles from the reported center of such a storm. The rapidity with which the weather can deteriorate with approach of the storm, and the violence of the fully developed tropical cyclone, are difficult to visualize if they have not been experienced.

Areas of Occurrence—Tropical cyclones occur almost entirely in six rather distinct regions, four in the northern hemisphere and two in the southern hemisphere, as shown in figure 10–1. The name by which such a disturbance is commonly known varies somewhat with the locality, as follows:

Region I. North Atlantic (West Indies, Caribbean Sea, Gulf of Mexico, and waters off the East Coast of the United States). A tropical cyclone with winds of 64 knots or greater is called a **hurricane.**

Region II. Southeastern North Pacific (waters off west coast of Mexico and Central America). The name **hurricane** is applied, as in Region I.

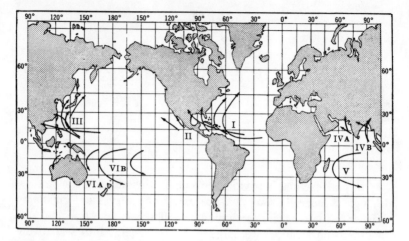

Figure 10–1 *Areas in which tropical cyclones occur, and their approximate tracks.*

Region III. Far East (the entire area west of the Mariana and Caroline Islands, across the Philippines and the China Sea, and northeastward to China and Japan). A fully developed storm with winds of 60 knots or greater is called a **typhoon** or, locally in the Philippine Islands, a **baguio.**

Region IV. A. Arabian Sea. *B.* Bay of Bengal. In these areas the storms are called **cyclones.**

Region V. South Indian Ocean (in the vicinity and to the east of Madagascar). As in Region IV, the tropical cyclone is called a **cyclone.**

Region VI. A. Australian waters (to longitude 160° E). *B.* South Pacific (the western portion, east of longitude 160° E). Several names are applied in this area, **cyclone** being the most common. One originating in the Timor Sea and moving southwest and then southeast across the interior of northwestern Australia is called a **willy-willy.** One to the east of Australia may be called a **hurricane.**

The only tropical ocean area in which tropical cyclones have not been encountered at some time is the South Atlantic.

As a tropical cyclone moves out of the tropics to higher latitudes, it normally loses energy slowly, expanding in area until it gradually dissipates or acquires the characteristics of extratropical cyclones. At any stage, a tropical cyclone normally loses energy at a much faster rate if it moves over land.

Season and Frequency of Occurrence—In Region III tropical cyclones may be encountered in any month of the year, though less frequently in winter than in summer. In the other regions, they occur only in the summer or autumn of that area, as shown in figure 10–2. The total number for the northern hemisphere reaches a sharp peak in September. In general, this is the month of greatest frequency in each of the first four regions, although the Far East reaches its maximum in August, and in the Arabian Sea there are two peaks, one in June, and the other in late October. In the southern hemisphere, the maximum number is not as sharply peaked, being distributed nearly equally over January, February, and March, the summer season of that hemisphere.

The occurrence of tropical cyclones in an area is not as regular as might be inferred from a curve such as any of those in figure 10–2 which are averages over a great many years. Even near the peak of a tropical cyclone season in any area there are periods when no tropical storms are observed. At the other extreme, as many as three hurricanes have been in progress at the same time in the North Atlantic, and as many as four typhoons in the Far East. The *average* total number of tropical cyclones occurring per year is 43 in the Northern Hemisphere and 13 in the Southern Hemisphere, or 56 throughout the world. However, the actual number in an area varies greatly from year to year. In the North Atlantic, where the greatest irregularity occurs, there have been as few as two and as many as 21 in a year, although the average number is seven. In the Far East, the number has varied from 13 to 25.

Storm Tracks—Tropical cyclones form over the ocean, in low latitudes. As one forms, it drifts slowly westward with the current of free air in which it forms. As it reaches the edge of a subtropical anticyclone, the storm, together with the general mass of air, drifts farther from the equator, in many instances curving poleward and then eastward with the winds of the general circulation. In general, a tropical cyclone moves very slowly at first, its speed varying from about five to 20 knots. The speed gradually increases as the storm progresses, and may, in a few instances, reach a value of 50 knots or more when the storm reaches temperate latitudes.

The average track varies somewhat as the season progresses, and individual storm tracks may differ widely from the average. Region I, the North Atlantic, is typical of the changes. In August, about 80 percent originate in the southern North Atlantic and the eastern Caribbean, and about 20 percent in the western Caribbean and Gulf of Mexico. About 60 percent curve toward the right, roughly parallel-

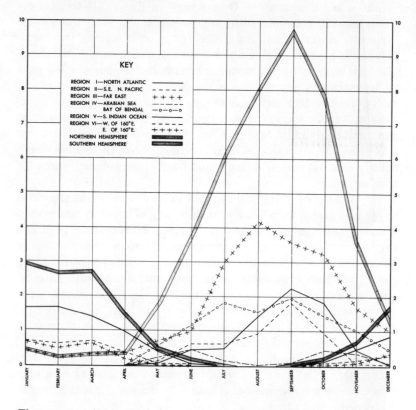

KEY

REGION I—NORTH ATLANTIC ————
REGION II—S.E. N. PACIFIC - - - - -
REGION III—FAR EAST + + + +
REGION IV—ARABIAN SEA - - - - -
         BAY OF BENGAL —o—o—o
REGION V—S. INDIAN OCEAN ————
REGION VI—W. OF 160°E. - - - - -
         E. OF 160°E. + + + -
NORTHERN HEMISPHERE
SOUTHERN HEMISPHERE

Figure 10–2 *Average number of tropical disturbances per month in the various regions.*

ing the coast of North America, and about 40 percent continue on westward, as shown in figure 10–3.

By the peak of the season, in September, the number forming in the southern North Atlantic and eastern Caribbean has dropped to 70 percent, but the number curving toward the right has increased to about the same percentage. The normal track has moved a little farther offshore in the lower latitudes, but has straightened somewhat so as to pass over eastern Newfoundland. This is shown in figure 10–4.

By October, the number originating in the southern North Atlantic and eastern Caribbean has dropped to 50 percent; and 80 percent of them curve, but at a point farther west, and more sharply, as shown

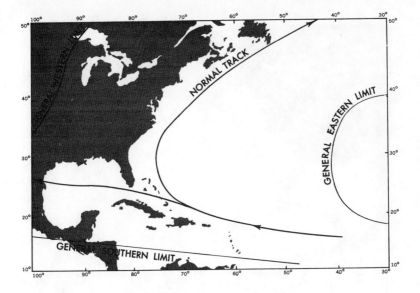

Figure 10–3 *Average North Atlantic storm tracks in August.*

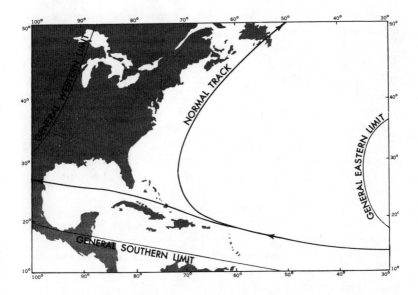

Figure 10–4 *Average North Atlantic storm tracks in September.*

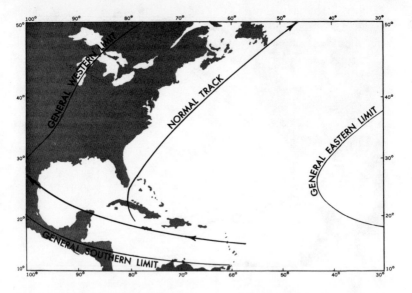

Figure 10–5 *Average North Atlantic storm tracks in October.*

in figure 10–5. By November, the change has been somewhat back toward the condition in September. As the season progresses, the deviation from average becomes greater and more common.

The differences, both in the averages and in individual tracks, are due to differences in the pressure pattern, particularly the location and movement of highs, which any cyclone tends to avoid.

Life Cycle—The life cycle of a tropical cyclone may be considered to consist of four rather distinct stages, as follows:

Formation.   A cyclonic circulation develops, and wind speed increases to hurricane force (64 knots) over a restricted area near the center. Atmospheric pressure drops to about 1,000 millibars (29.53 inches). This stage may occupy several days, or may be completed in a period of 12 hours or less.

Immaturity.   The pressure at the center continues to fall (the storm "deepens") and the wind speed increases, but the storm is still confined to a small area.

Maturity.   The pressure at the center remains about the same, but the area of hurricane winds expands, to a radius of perhaps 150 to 200 miles, with winds of gale force extending to perhaps 300 miles. Individual storms may differ considerably from these averages.

Decay. The area continues to increase, the pressure at the center rises, and the wind speed decreases. The storm loses its tropical characteristics and gradually dissipates, a process that may require several days over an ocean area. Over land the decay is more rapid.

Origin and Development—Tropical cyclones originate between the doldrums and the zones of the strongest trade winds. This accounts, at least partly, for the absence of such storms in the South Atlantic, for the Atlantic doldrums remain several degrees north of the equator except for occasional brief periods.

Some of the details regarding the formation of tropical cyclones are not understood, but the fact that such storms form only over water, and dissipate rapidly if they encounter land, probably indicates the need for a supply of water vapor. Over the tropical ocean this is abundantly available in the lower portion of the atmosphere. When a low develops over tropical oceans, hot, vapor-laden air flows in from adjacent regions. This air ascends near the center of the low, and condensation occurs. Each pound of water vapor that condenses into cloud or rain liberates approximately 970 British thermal units of heat. This heat warms the surrounding air, thus increasing further the instability, and hastening the ascent of the air. Thus, the pressure continues to drop and the winds increase in speed, bringing in an increasing quantity of warm, moist air from the regions surrounding the low. At some height, the ascending air flows outward from the center of the cyclonic circulation. This process of inward flow, rising air current, condensation, warming, and high-level outflow causes the low to deepen and the wind speed to increase. Thus, as long as conditions remain suitable, the storm grows more intense.

While the actual mechanics of tropical cyclone formation are somewhat more involved than just described, the essential steps are given. Several theories exist regarding the details of initial formation of the low pressure area. Dropping pressure at the surface due to disturbances at high levels of the atmosphere; interaction of two air streams to produce a cyclonic eddy, causing convergence of surface air and the resulting ascent; and the joining of minor disturbances in the wind and pressure patterns in the atmosphere are all considered possibilities. The process is probably begun by several factors which combine in just the right relationship.

When it becomes fully developed, a tropical cyclone covers a well-defined area, more or less circular in shape, within which the atmospheric pressure decreases rapidly toward the center. This decrease of pressure may amount to a maximum of 0.01 or even 0.02 inch per mile. Because of the rapid decrease of pressure with distance, the wind speed is high, being greatest at the regions of steepest pressure gradient.

At the center of the storm, there is normally an area five to 30 miles in diameter (most often ten to 15 miles) within which the wind speed drops to a relative calm, usually ten to 15 knots or less. This is the **eye of the storm.** Ascending air causes the dense cover of clouds to give way to a thin layer of low clouds with holes through which the sun may shine. Around the edge of the eye, the wind speed increases from the relative calm to the full fury of maximum speed within a distance of a few feet. Here the heavy cloud seems thickest, and the torrential rains surrounding the central area appear concentrated. This is the **wall of the eye.** When a tropical cyclone moves to higher latitude, its eye becomes less clearly defined as the maximum wind moves outward from the center, the wall of the eye becomes more indistinct, and its cloud cover increases.

Locating and Tracking a Tropical Cyclone—By means of radio, organized meteorological services collect weather observations daily from island stations, ships at sea, and aircraft. When a tropical cyclone is located, usually in its early formative stage, it is followed closely. In the North Atlantic, aircraft of the U. S. Navy and U. S. Air Force, in cooperation with the Weather Bureau, make frequent flights to the vicinity of such storms to provide information needed for tracking the hurricane and determining its intensity. Bulletins are broadcast to ships several times daily, giving information on each storm's location, intensity, and movement. As a further aid, the mariner may obtain weather reports by radio directly from other ships in the vicinity of a tropical cyclone. Radar may be used to follow the movements of the precipitation areas when they are within range.

Although these aids normally prove adequate for locating and avoiding a tropical cyclone, knowledge of the appearance of the sea and sky in the vicinity of such a storm is useful to the mariner.

The passage of a tropical cyclone at sea is an experience not soon to be forgotten.

An early indication of the approach of such a storm is the presence of a long swell. In the absence of a tropical cyclone, the crests of swell in the deep waters of the Atlantic pass at the rate of perhaps eight per minute. Swell generated by a hurricane is about twice as long, the crests passing at the rate of perhaps four per minute. Swell may be observed several days before arrival of the storm.

When the storm center is 500 to 1,000 miles away, the barometer usually rises a little, and the skies are relatively clear. Cumulus clouds, if present at all, are few in number and their vertical development appears suppressed. The barometer usually appears restless, **pumping** up and down a few hundreths of an inch.

As the tropical cyclone comes nearer, a cloud sequence begins which resembles that associated with the approach of a warm front in middle latitudes. Snow-white, fibrous "mare's tails" (cirrus) appear when the storm is about 300 to 600 miles away. Usually these seem to converge, more or less, in the direction from which the storm is approaching. This convergence is particularly apparent at about the time of sunrise and sunset.

Shortly after the cirrus appears, but sometimes before, the barometer starts a long, slow fall. At first the fall is so gradual that it only appears to alter somewhat the normal daily cycle (two maxima and two minima in the tropics). As the rate of fall increases, the daily pattern is completely lost in the more or less steady fall.

The cirrus becomes more confused and tangled, and then gradually gives way to a continuous veil of cirrostratus. Below this veil, altostratus forms, and then stratocumulus. These clouds gradually become more dense, and as they do so, the weather becomes unsettled. A fine, mist-like rain begins to fall, interrupted from time to time by showers. The barometer has fallen perhaps a tenth of an inch.

As the fall becomes more rapid, the wind increases in gustiness, and its speed becomes greater, reaching a value of perhaps 22 to 40 knots (Beaufort 6–8). On the horizon appears a dark wall of heavy cumulonimbus, the **bar** of the storm. Portions of this heavy cloud become detached from time to time and drift across the sky, accompanied by rain squalls and wind of increasing speed. Between squalls, the cirrostratus can be seen through breaks in the stratocumulus.

As the bar approaches, the barometer falls more rapidly and wind speed increases. The seas, which have been gradually mounting, become tempestuous. Squall lines, one after the other, sweep past in ever increasing number and intensity.

With the arrival of the bar, the day becomes very dark, squalls become virtually continuous, and the barometer falls precipitously, with a rapid increase in wind speed. The center may still be 100 to 200 miles away in a fully developed tropical cyclone. As the center of the storm comes closer, the ever-stronger wind shrieks through the rigging and about the superstructure of the vessel. As the center approaches, rain falls in torrents. The wind fury increases. The seas be-

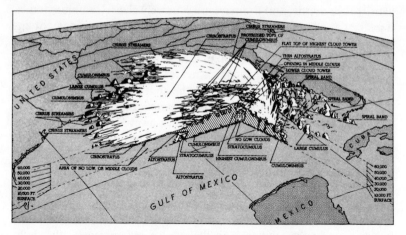

Figure 10–6 *Typical hurricane cloud formations.*

come mountainous. The tops of huge waves are blown off to mingle with the rain and fill the air with water. Objects at a short distance are not visible. Even the largest and most seaworthy vessels become virtually unmanageable, and may sustain heavy damage. Less sturdy vessels do not survive. Navigation virtually stops as safety of the vessel becomes the prime consideration. The awesome fury of this condition can only be experienced. Words are inadequate to describe it.

If the eye of the storm passes over the vessel, the winds suddenly drop to a breeze as the wall of the eye passes. The rain stops, and the skies clear sufficiently to permit the sun to shine through holes in the comparatively thin cloud cover. Visibility improves. Mountainous seas approach from all sides, apparently in complete confusion. The barometer reaches its lowest point, which may be an inch and a half or two inches below normal in fully developed tropical cyclones. As the wall on the opposite side of the eye arrives, the full fury of the wind strikes as suddenly as it ceased, but from the opposite direction. The sequence of conditions that occurred during approach of the storm is reversed, and pass more quickly, as the various parts of the storm are not as wide in the rear of a storm as on its forward side.

Typical cloud formations associated with a hurricane are shown in figure 10–6.

Locating the Center of a Tropical Cyclone—If intelligent action is to be taken to avoid the full fury of a tropical cyclone, early determination of its location and direction of travel relative to the vessel is essential. The bulletins and forecasts are an excellent general guide, but they are not infallible and may be sufficiently in error to induce a mariner in a critical position to alter course so as to unwittingly increase the danger to his vessel. Often it is possible, using only those observations made aboard ship, to obtain a sufficiently close approximation to enable the vessel to maneuver to the best advantage.

As stated earlier, the presence of an exceptionally long swell is usually the first visible indication of the existence of a tropical cyclone. In deep water it approaches from the general direction of origin (the position of the storm center *when the swell was generated*). However, in shoaling water this is a less reliable indication because the direction is changed by refraction, the crests being more nearly parallel to the bottom contours.

When the cirrus clouds appear, their point of convergence provides an indication of the direction of the storm center. If the storm is to pass well to one side of the observer, the point of convergence shifts slowly in the direction of storm movement. If the storm center will pass near the observer, this point remains steady. When the bar becomes visible, it appears to rest upon the horizon for several hours. The darkest part of this cloud is in the direction of the storm center. If the storm is to pass to one side, the bar appears to drift slowly along the horizon. If the storm is heading directly toward the observer, the position of the bar remains fixed. Once within the area of the dense, low clouds, one should observe their direction of movement, which is almost exactly along the isobars, with the center of the storm being 90° from the direction of cloud movement (left of direction of movement in the northern hemisphere, and right in the southern hemisphere).

The winds are probably the best guide to the direction of the center of a tropical cyclone. The circulation is cyclonic, but because of the steep pressure gradient near the center, the winds there blow with greater violence and are more nearly circular than in extratropical cyclones.

According to Buys Ballot's law an observer who faces into the wind has the center of the low pressure on his right in the northern hemisphere, and on his left in the southern hemisphere, and in each case somewhat behind him. If the wind followed circular isobars exactly, the center would be exactly eight points, or 90°, from dead ahead when facing into the wind. However, the track of the wind is usually inclined somewhat toward the center, so that the angle from dead ahead varies between perhaps 8 and 12 points (90° to 135°). The inclination varies in different parts of the same storm. It is least in front of the storm, and greatest in the rear, since the actual wind is the vector sum of that due to the pressure gradient and the motion of the storm along the track. A good average is perhaps ten points in front, and 11 or 12 points in the rear. These values apply when the storm center is still several hundred miles away. Closer to the center, the wind blows more nearly along the isobars, the inclination being reduced by one or two points at the wall of the eye. Since wind direction usually shifts temporarily during a squall, its direction at this time should not be used for determining the position of the center. The approximate relationship of wind to isobars and storm center in the northern hemisphere is shown in figure 10–7.

When the center is within radar range, it might be located by this equipment. However, since the radar return is predominantly from the rain, results can be deceptive, and other indications should not be neglected. A typical radar PPI presentation of a tropical cyclone is shown in figure 10–8.

Distance from the storm center is more difficult to determine than direction. Radar is perhaps the best guide. However, the rate of fall of the barometer is some indication. If a vessel is hove-to in front of a storm which is advancing directly toward it, the fall of pressure per hour might be about that shown in figure 10–9. However, this is an imperfect indication, for the rate of fall may be quite erratic, and will vary somewhat with the depth of the low at the center, the speed of the storm center along its track, and the stage in the life cycle of the storm. The usefulness of this information is further reduced by the fact that a vessel would not normally remain hove-to in the path of a tropical cyclone.

Maneuvering to Avoid the Storm Center—The safest procedure with respect to tropical cyclones is to avoid them. If action is taken sufficiently early, this is simply a matter of setting a course that will take the vessel well to one side of the probable track of the storm, and then continuing to plot the positions of the storm center, as given in the weather bulletins, revising the course as needed.

However, such action is not always possible. If one finds himself within the storm area, the proper action to take depends in part upon his position relative to the storm center and its direction of travel. It is customary to divide the circular area of the storm into two parts. In the northern hemisphere, that part to the *right* of the storm track

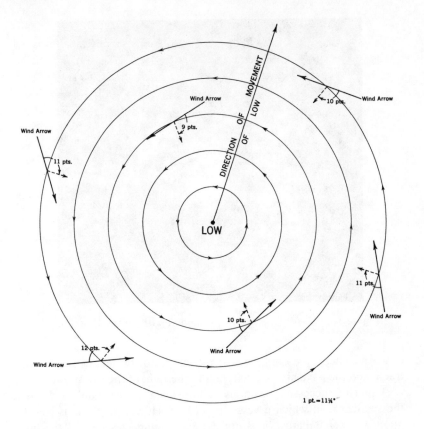

Figure 10–7 *Approximate relationship of wind to isobars and storm center in the northern hemisphere.*

(facing in the direction *toward* which the storm is moving) is called the **dangerous semicircle.** It is considered dangerous because (1) the actual wind *speed* is greater than that due to the pressure gradient alone, since it is augmented by the forward motion of the storm, and (2) the *direction* of the wind and sea is such as to carry a vessel into the path of the storm (in the forward part of the semicircle). The part to the left of the storm track is called the **navigable semicircle.** In this part, the wind is decreased by the forward motion of the storm, and the wind blows vessels away from the storm track (in the forward part). Because of the greater wind speed in the dangerous semicircle, the seas are higher here than in the navigable semicircle. In the south-

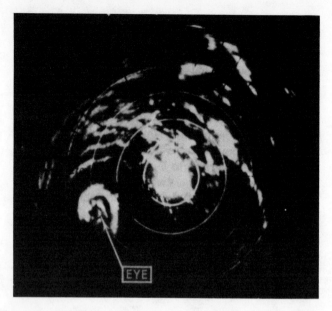

Figure 10–8 *Typical radar PPI presentation of a tropical cyclone.*

ern hemisphere, the dangerous semicircle is to the left of the storm track, and the navigable semicircle is to the right of the storm track.

A plot of successive positions of the storm center should indicate the semicircle in which a vessel is located. However, if this is based upon weather bulletins, it is not a reliable guide because of the lag between the observations upon which the bulletin is based and the time of reception of the bulletin, with the ever present possibility of a change in the direction of motion of the storm. The use of one's radar eliminates this lag, but the return is not always a true indication of the center. Perhaps the most reliable guide is the wind. Within the cyclonic circulation, a *veering* wind (one changing direction to the right in the northern hemisphere and to the left in the southern hemisphere) indicates a position in the dangerous semicircle, and a *backing* wind (one changing in a direction opposite to a veering wind) indicates a position in the navigable semicircle. However, if a vessel is underway, its motion should be considered. If it is outrunning the storm or pulling rapidly toward one side (which is not difficult during the early stages of a storm, when its speed is low), the opposite effect occurs. This should usually be accompanied by a rise in atmospheric

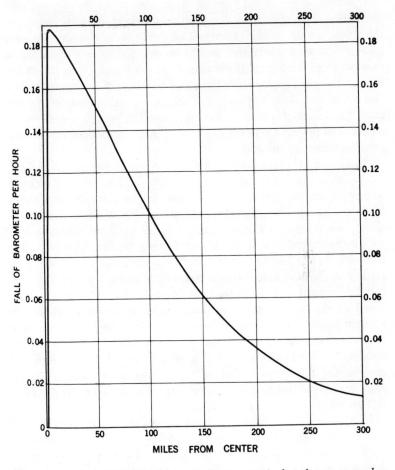

Figure 10–9 *Typical average pressure drop as tropical cyclone approaches.*

pressure, but if motion of the vessel is nearly along an isobar, this may not be a reliable indication. If in doubt, the safest action is usually to stop long enough to determine definitely the semicircle. The loss in valuable time may be more than offset by the minimizing of the possibility of taking the wrong action and increasing the danger to the vessel. If the wind direction remains steady (for a vessel which is stopped), with increasing speed and falling barometer, the vessel is in or near the path of the storm. If it remains steady with decreasing speed and rising barometer, the vessel is on the storm track, behind the center.

The first action to take if one finds himself within the cyclonic circulation, is to determine the position of his vessel with respect to the storm center. While the vessel can still make considerable way through the water, a course should be selected to take it as far as possible from the center. If the vessel can move faster than the storm, it is a relatively simple matter to outrun the storm if sea room permits. But when the storm is faster, the solution is not as simple. In this case, the vessel, if ahead of the storm, will approach nearer to the center. The problem is to select a course that will produce the greatest possible minimum distance.

As a very general rule, for a vessel in the northern hemisphere, safety lies in placing the wind on the starboard bow in the dangerous semicircle and on the starboard quarter in the navigable semicircle. If on the storm track ahead of the storm, the wind should be put about two points on the starboard quarter until the vessel is well within the navigable semicircle, and the rule for that semicircle then followed. In the southern hemisphere the same rules hold, but with respect to the port side. With a faster than average vessel, the wind can be brought a little farther aft in each case. However, as the speed of the storm increases along its track, the wind should be brought farther forward. If land interferes with what would otherwise be the best maneuver, the solution should be altered to fit the circumstances. If the speed of a vessel is greater than that of the storm, it is possible for the vessel, if behind the storm, to overtake it. In this case, the only action usually needed is to slow enough to let the storm pull ahead.

In all cases, one should be alert to changes in the direction of movement of the storm center, particularly in the area where the track normally curves toward the pole. If the storm maintains its direction and speed, the ship's course should be maintained as the wind shifts.

If it becomes necessary for a vessel to heave to, the characteristics of the vessel should be considered. A power vessel is concerned primarily with damage by direct action of the sea. A good general rule is to heave to with head to the sea in the dangerous semicircle or stern to the sea in the navigable semicircle. This will result in greatest amount of headway away from the storm center, and least amount of leeway toward it. If a vessel handles better with the sea astern or on the quarter, it may be placed in this position in the navigable semicircle or in the rear half of the dangerous semicircle, but *never* in the forward half of the dangerous semicircle. It has been reported that when the wind reaches hurricane speed and the seas become confused, some ships ride out the storm best if the engines are stopped, and the vessel is permitted to seek its own position. In this way, it is said, the ship rides *with* the storm instead of fighting *against* it.

In a sailing vessel, while attempting to avoid a storm center, one should steer courses as near as possible to those prescribed above for power vessels. However, if it becomes necessary for such a vessel to heave to, the wind is of greater concern than the sea. A good general rule always is to heave to on whichever tack permits the shifting wind to draw aft. In the northern hemisphere this is the starboard tack in the dangerous semicircle and the port tack in the navigable semicircle. In the southern hemisphere these are reversed.

While each storm requires its own analysis, and frequent or continual resurvey of the situation, the general rules for a steamer may be summarized as follows:

Northern Hemisphere

Right or Dangerous Semicircle—Bring the wind on the starboard bow (045° relative), hold course and make as much way as possible. If obliged to heave to, do so with head to the sea.

Left or Navigable Semicircle—Bring the wind on the starboard quarter (135° relative), hold course and make as much way as possible. If obliged to heave to, do so with stern to the sea.

On Storm Track, Ahead of Center—Bring the wind two points on the starboard quarter (157°.5 relative), hold course and make as much way as possible. When well within the navigable semicircle, maneuver as indicated above.

On Storm Track, Behind Center—Avoid the center by the best practicable course, keeping in mind the tendency of tropical cyclones to curve northward and eastward.

Southern Hemisphere

Left or Dangerous Semicircle—Bring the wind on the port bow (315° relative), hold course and make as much way as possible. If obliged to heave to, do so with head to the sea.

Right or Navigable Semicircle—Bring the wind on the port quarter (225° relative), hold course and make as much way as possible. If obliged to heave to, do so with stern to the sea.

On Storm Track, Ahead of Center—Bring the wind two points on the port quarter (202°.5 relative), hold course and make as much way as possible. When well within the navigable semicircle, maneuver as indicated above.

On Storm Track, Behind Center—Avoid the center by the best practicable course, keeping in mind the tendency of tropical cyclones to curve southward and eastward.

Whenever a tropical cyclone is encountered, the wise procedure is to begin preparing the vessel for heavy weather in sufficient time to permit thorough preparation, so that damage may be minimized. One should be particularly careful to keep free surfaces of liquids to a minimum.

Coastal Effects—The high winds of a tropical cyclone inflict widespread damage when such a storm leaves the ocean and crosses land. Aids to navigation may be blown out of position or destroyed. Craft in harbors, unless they are properly secured, drag anchor or are blown against obstructions. Ashore, trees are blown over, houses are damaged, power lines are blown down, etc. The greatest damage usually occurs in the dangerous semicircle a short distance from the center, where the strongest winds occur. As the storm continues on across land, its fury subsides faster than it would if it had remained over water.

Along the coast, particularly, greater damage may be inflicted by water than by the wind. There are at least four sources of water damage. First, the unusually high seas generated by the storm winds pound against shore installations and craft in their way. Second, the continued blowing of the wind toward land causes the water level to increase perhaps three to ten feet above its normal level. This **storm tide,** which may begin when the storm center is 500 miles or even farther from the shore, gradually increases until the storm passes. The highest storm tides are caused by a slow-moving tropical cyclone of large diameter, because both of these effects result in greater duration of wind in the same direction. The effect is greatest in a partly enclosed body of water, such as the Gulf of Mexico, where the concave coastline does not readily permit the escape of water. It is least on small islands, which present little obstruction to the flow of water. Third, the furious winds which blow around the wall of the eye create a ridge of water called a **storm wave,** which strikes the coast and often inflicts heavy damage. The effect is similar to that of a **seismic sea wave,** caused by an earthquake in the ocean floor. Both of these waves are popularly called **tidal waves.** Storm waves of 20 feet or more have occurred. About three or four feet of this is due to the decrease of atmospheric pressure, and the rest to winds. Like the damage caused by wind, that due to high seas, the storm tide, and the storm wave is greatest in the dangerous semicircle, near the center. The fourth source of water damage is the heavy rain that accompanies a tropical cyclone. This causes floods that add to the damage caused in other ways.

When proceeding along a shore recently visited by a tropical cyclone, a navigator should remember that time is required to restore aids to navigation which have been blown out of position or destroyed. In some instances the aid may remain but its light, sound apparatus, or radiobeacon may be inoperative. Landmarks may have been damaged or destroyed.

Appendix A
Glossary

This appendix is an abridgment of the definitions of H.O. Pub. No. 220, *Navigation Dictionary*.

absolute humidity.—The mass of water vapor per unit of volume of air.

absolute zero.—The lowest possible temperature, about $(-)459°.67$ for $(-)273°.15C$.

air mass.—An extensive body of air within which the conditions of temperature and moisture in a horizontal plane are essentially uniform.

air temperature correction.—That sextant altitude correction due to changes in refraction caused by difference between the actual temperature and the standard temperature used in the computation of the refraction table.

altocumulus.—A cloud layer (or patches) within the middle level (mean height 6,500—20,000 ft.) composed of rather flattened globular masses, the smallest elements of the regularly arranged layers being fairly thin, with or without shading. These elements are arranged in groups, in lines or in waves, following one or two directions, and are sometimes so close together that their edges join.

altostratus.—A sheet of gray or bluish cloud within the middle level (mean height 6,500—20,000 ft.). Sometimes the sheet is composed of a compact mass of dark, thick, gray clouds of fibrous structure; at other times the sheet is thin and through it the sun or moon can be seen dimly as though gleaming through ground glass.

anemometer.—An instrument for measuring the speed of the wind. Some instruments also indicate the direction from which it is blowing.

aneroid barometer.—An instrument which determines atmospheric pressure by the effect of such pressure on a thin-metal cylinder from which the air has been partly exhausted.

annular eclipse.—An eclipse in which a thin ring of the source of light appears around the obscuring body.

anticyclone.—An approximately circular portion of the atmosphere, having relatively high atmospheric pressure and winds which blow clockwise around the center in the northern hemisphere and counterclockwise in the southern hemisphere.

apogean tides.—Tides of decreased range occurring when the moon is near apogee.

apogee.—That orbital point farthest from the earth when the earth is the center of attraction (as in the case of the moon).

apparent horizon.—Visible horizon.

apparent motion.—Motion relative to a specified or implied reference point which may itself be in motion.

apparent sun.—The actual sun as it appears in the sky.

apparent wind.—Wind relative to a moving point, such as a vessel.

arming.—Tallow or other substance placed in the recess at the lower end of a sounding lead, for obtaining a sample of the bottom.

astronomical tide.—Tide related to the attractions of celestial bodies particularly the sun and moon.

astronomical twilight.—The period of incomplete darkness when the upper limb of the sun is below the visible horizon, and the center of the sun is not more than 18° below the celestial horizon.

atmosphere.—The envelope of air surrounding the earth or other celestial body.

atmospheric absorption.—The loss of power in transmission of radiant energy by dissipation in the atmosphere.

atmospheric noise.—Static.

atmospheric pressure.—The pressure exerted by the weight of the earth's atmosphere. Its standard value at sea level is about 14.7 pounds per square inch.

attenuation.—A lessening in amount, particularly the reduction of the amplitude of a wave with distance from the origin.

aurora.—A luminous phenomena due to electrical discharge in the upper atmosphere, most commonly seen in high latitudes.

aurora australis.—The aurora in the southern hemisphere.

aurora borealis.—The aurora in the northern hemisphere.

auroral zone.—The area of maximum auroral activity.

autumnal equinox.—That point of intersection of the ecliptic and the celestial equator occupied by the sun as it changes from north to south declination, on or about September 23, or the instant this occurs.

awash.—Situated so that the top is intermittently washed by waves or tidal action.

back.—Of the wind, to change direction counterclockwise in the northern hemisphere and clockwise in the southern hemisphere.

barograph.—A recording barometer.

barometer.—An instrument for measuring atmospheric pressure.

barometric pressure.—Atmospheric pressure as indicated by a barometer.

barometric tendency.—The change of barometric pressure within a specified time (usually three hours) before an observation, together with the direction of change and the characteristics of the rise or fall.

bathymetric chart.—A topographic chart of the bed of a body of water.

bathythermograph.—A recording thermometer for determining temperature of the sea at various depths.

Beaufort scale.—A numerical scale for indicating wind speed, named after Admiral Sir Francis Beaufort, who devised it in 1806.

beset.—Surrounded so closely by sea ice that steering control is lost.

bottom sample.—A portion of the material forming the bottom, brought up for inspection.

cable.—1. A unit of distance equal to 720 feet in the U.S. Navy. 2. A chain or strong fiber or wire rope used to anchor or moor vessels or buoys. 3. A stranded electric conductor or several conductors laid up together.

calving.—The breaking away of a mass of ice from a parent iceberg, glacier, or ice shelf.

Celsius temperature.—Temperature based upon a scale in which, under standard atmospheric pressure, water freezes at 0° and boils at 100°. Called "centigrade temperature" before 1948.

centigrade temperature.—Celsius temperature.

change of tide.—A reversal of the direction of motion (rising or falling) of a tide.

character of the bottom.—The type of material of which the bottom is composed.

chart.—A map intended primarily for navigational use.

chart datum.—The tidal datum to which soundings on a chart are referred.

charted depth.—The vertical distance from the chart datum to the bottom.

cirrocumulus.—High clouds (mean lower level above 20,000 ft.) composed of small white flakes or of very small globular masses, usually without shadows, which are arranged in groups or lines, or more often in ripples resembling those of sand on the seashore.

cirrostratus.—Thin, whitish, high clouds (mean lower level above 20,000 ft.) sometimes covering the sky completely and giving it a milky appearance and at other times presenting, more or less distinctly, a formation like a tangled web.

cirrus.—Detached high clouds (mean lower level above 20,000 ft.) of delicate and fibrous appearance, without shading, generally white in color, and often of a silky appearance.

civil twilight.—The period of incomplete darkness when the upper limb of the sun is below the visible horizon, and the center of the sun is not more than 6° below the celestial horizon.

coastal current.—An ocean current flowing roughly parallel to a coast, outside the surf zone.

cold air mass.—An air mass that is colder than surrounding air, and usually colder than the surface over which it is moving.

cold front.—That line of discontinuity, at the earth's surface or at a horizontal plane aloft, along which an advancing cold air mass is undermining and displacing a warmer air mass.

continuous wave.—A series of waves of like amplitude and frequency.

contour.—A line connecting points of equal elevation or equal depth.

controlling depth.—The least depth in the approach or channel to an area, such as a port, governing the maximum draft of vessels that can enter.

Coriolis force.—An apparent force acting on a body in motion, due to rotation of the earth, causing deflection to the right in the northern hemisphere and to the left in the southern hemisphere.

countercurrent.—A secondary current flowing adjacent and in the opposite direction to another current.

cumulonimbus.—A massive cloud with great vertical development, the summits of which rise in the form of mountains or towers, the upper parts often spreading out in the form of an anvil.

cumulus.—A dense cloud with vertical development, having a horizontal base and dome-shaped upper surface, exhibiting protuberances.

current.—1. Water in essentially horizontal motion. 2. A hypothetical horizontal motion of such set and drift as to account for the difference between a dead reckoning position and a fix at the same time. 3. Air in essentially vertical motion. 4. Electricity flowing along a conductor.

current chart.—A chart on which current data are graphically depicted.

current diagram.—A graph showing the average speeds of flood and ebb currents throughout the current cycle for a considerable part of a tidal waterway.

current difference.—The difference between the time of slack water or strength of current at a subordinate station and at its reference station.

current direction.—The direction *toward* which a current is flowing.

current meter.—An instrument for measuring the speed of a current, and sometimes the direction of flow.

current rips.—Small waves formed on the surface of water by the meeting of opposing ocean currents.

cyclone.—An approximately circular portion of the atmosphere, having relatively low atmospheric pressure and winds which blow counterclockwise around the center in the northern hemisphere and clockwise in the southern hemisphere.

damping.—The progressive diminishing of amplitude of oscillations, waves, etc.

deep.—An unmarked fathom point on a lead line.

deep sea lead (lĕd).—A heavy sounding lead (about 30 to 100 pounds), usually having a line 100 fathoms or more in length.

depth.—Vertical distance from a given water level to the bottom.

depth contour.—A contour connecting points of equal depth.

dew point.—The temperature to which air must be cooled at constant pressure and constant water vapor content to reach saturation.

diaphone.—A device for producing a distinctive fog signal by means of a slotted reciprocating piston actuated by compressed air.

direction of current.—The direction *toward* which a current is flowing.

direction of waves or swell.—The direction *from* which waves or swell are moving.

direction of wind.—The direction *from* which a wind is blowing.

diurnal current.—Tidal current having one flood current and one ebb current each tidal day.

diurnal inequality.—The difference between the heights of the two high tides or two low tides during the tidal day, or the difference in speed between the two flood currents or the two ebb currents during a tidal day.

diurnal tide.—Tide having one high tide and one low tide each tidal day.

double tide.—A high tide consisting of two maxima of nearly the same height separated by a relatively small depression, or a low tide consisting of two minima separated by a relatively small elevation.

drift.—1. The speed of a current. 2. The distance a vessel is moved by current and wind. 3. Downwind or down current motion due to wind or current.

drift current.—Any broad, shallow, slow-moving ocean current.

drogue.—Sea anchor.

ebb current. Tidal current moving away from land or down a tidal stream.

equatorial tides.—The tides that occur when the moon is near the celestial equator, when the difference in height between consecutive high or low tides is a minimum.

equinoctial tides.—The tides that occur at or about the time of the equinoxes, when the spring range is greater than average.

equinox.—One of the two points of intersection of ecliptic and the celestial equator, or the instant the sun occupies one of these points, when its declination is 0°.

eye of the storm.—The center of a tropical cyclone.

Fahrenheit temperature.—Temperature based upon a scale in which, under standard atmospheric pressure, water freezes at 32° and boils at 212°.

fair tide.—A tidal current which increases the speed of a vessel.

fair wind.—A wind which aids a craft in making progress in a desired direction.

falling tide.—A tide in which the depth of water is decreasing.

fathom.—A unit of length equal to six feet.

fathom curve, fathom line.—A depth contour with depth measured in fathoms.

Fathometer.—The trade name for a widely used echo sounder.

favorable current.—A current which increases the speed of a vessel over the ground.

favorable wind.—A wind which helps a craft make progress in a desired direction.

feel the bottom.—The action of a vessel in shoal water, when its speed is reduced and it sometimes becomes hard to steer.

floe.—Sea ice, either a single unbroken piece or many individual pieces, covering an area of water.

floeberg.—A mass of heavily hummocked sea ice resembling an iceberg in appearance.

flood current.—Tidal current moving toward land or up a tidal stream.

fog.—A visible assemblage of numerous tiny droplets of water, or ice crystals formed by condensation of water vapor in the air, with the base at the surface of the earth.

fog signal.—A warning signal transmitted by a vessel or aid to navigation during periods of low visibility.

front.—The intersection of a frontal surface and a horizontal plane.

frontal surface.—The thin zone of discontinuity between two air masses.

frost smoke.—Fog produced by apparent steaming of a relatively warm sea in the presence of much colder air.

glacier.—A field or stream of ice which moves or has moved slowly down an incline.

gradient tints.—A series of color tints used on some charts to indicate relative heights or depths.

greater ebb.—The stronger of two ebb currents occurring during a tidal day.

greater flood.—The stronger of two flood currents occurring during a tidal day.

ground swell.—A long ocean wave, or series of waves, in shoal water, at a considerable distance from its origin.

growler.—A small iceberg, piece broken from an iceberg, or detached piece of sea ice, large enough to be a hazard to shipping but small enough that it may escape detection.

half-tide level.—The level midway between mean high water and mean low water.

haul.—Of the wind, to shift in a counterclockwise direction, or to shift forward of a vessel.

haze.—Fine dust or salt particles in the air, too small to be individually apparent but in sufficient number to reduce visibility and cast a bluish or yellowish veil over the landscape, subduing its colors.

height of tide.—Vertical distance from the tidal datum to the level of the water at any time.

higher high water.—The higher of two high tides occurring during a tidal day.

higher low water.—The higher of two low tides occurring during a tidal day.

high tide.—The maximum height reached by a rising tide.

high water.—High tide.

high water full and change.—The average interval of time between the transit (upper or lower) of the full or new moon and the next high water.

high water inequality.—The difference between the height of the two high tides during a tidal day.

high water lunitidal interval.—The interval of time between the transit (upper or lower) of the moon and the next high water at a place.

humidity.—The amount of water vapor in the air.

hummock.—A mound or hill in pressure ice.

hydrographic survey.—A survey of a water area.

hydrography.—That science which deals with the measurement of the physical features of waters and their marginal land areas, with special reference to the elements that affect safe navigation, and the publication of such information in a form suitable for use of navigators.

hydrolant.—An urgent notice of dangers to navigation in the Atlantic.

hydrometeor.—Any product from the condensation of atmospheric water vapor, whether formed in the free atmosphere or at the earth's surface.

hydropac.—An urgent notice of dangers to navigation in the Pacific.

hygrometer.—An instrument for measuring the humidity of the air.

ice barrier.—Impenetrable ice.

iceberg.—A mass of land ice which has broken away from its parent formation on the coast and either floats in the sea or is stranded.

ice chart.—A chart showing prevalence of ice, usually with reference to navigable waters.

ice field.—Sea ice covering an area greater than five miles across.

ice jam.—An accumulation of broken ice caught in a narrow part of a stream or blown against the shore of a lake.

ice shelf.—A thick ice formation with level surface extending over the sea but attached to the land.

ice tongue.—A narrow peninsula of ice.

kilometer.—One thousand meters (about 0.54 nautical mile).

knot.—A unit of speed equal to one nautical mile per hour.

land ice.—All ice formed on land.

land mile.—Statute mile.

lee.—That side toward which the wind blows.

leeway.—The leeward motion of a vessel, due to wind, expressed as distance, speed, or an angle.

lesser ebb.—The weaker of two ebb currents occurring during a tidal day.

lesser flood.—The weaker of two flood currents occurring during a tidal day.

line of soundings.—A series of soundings obtained by a vessel underway, usually at regular intervals.

loom.—The glow of a light which is below the horizon, caused by reflection by solid particles in the air.

lower high water.—The lower of two high tides occurring during a tidal day.

low tide.—The minimum height reached by a falling tide.

low water.—Low tide.

low water inequality.—The difference between the heights of the two low tides during a tidal day.

low water lunitidal interval.—The interval of time between the transit (upper or lower) of the moon and the next low water at a place.

lunar tide.—That part of the tide due solely to the tide-producing force of the moon.

lunitidal interval.—The interval of time between the transit (upper or lower) of the moon and the next high water or low water at a place.

maneuvering board.—A polar coordinate plotting sheet devised to facilitate solution of problems involving relative movement.

map.—A representation, usually on a plane surface, of all or part of the surface of the earth, celestial sphere, or other area; showing relative size and position, according to a given projection, of the various features represented.

March equinox.—Vernal equinox.

marine navigation.—The navigation of water craft.

maximum ebb.—The greatest speed of an ebb current.

maximum flood.—The greatest speed of a flood current.

mean sea level.—The average height of the surface of the sea for all stages of the tide, usually determined from hourly readings.

mean tide level.—Half-tide level.

mercurial barometer.—An instrument which determines atmospheric pressure by measuring the height of a column of mercury which the atmosphere will support.

meterological tide.—A change in water level due to meterological conditions.

meterology.—The science of the atmosphere.

millibar.—A unit of pressure equal to 1,000 dynes per square centimeter.

mist.—Thin fog of relatively large particles, or very fine rain.

mixed current.—A type of tidal current characterized by a conspicuous difference in speed between the two flood currents or two ebb currents usually occurring each tidal day.

mixed tide.—A type of tide having a large inequality in the heights of either the two high tides or the two low tides usually occurring each tidal day.

nautical chart.—A chart intended primarily for marine navigation.

nautical mile.—A unit of distance equal to 1,852 meters (6,076.11549 U.S. feet, approximately). This is equal approximately to the length of 1′ of latitude.

navigable semicircle.—That half of a cyclonic storm area to the left of the storm track in the northern hemisphere, and to the right of the storm track in the southern hemisphere. In this semicircle the winds are weaker and tend to blow a vessel away from the path of the storm.

navigation.—The process of directing the movement of a craft from one point to another.

neap tides.—The tides occurring near the times of first and last quarter of the moon, when the range of tide tends to decrease.

nimbostratus.—A dark, low shapeless cloud layer (mean upper level below 6,500 ft.) usually nearly uniform; the typical rain cloud.

nimbus.—A characteristic rain cloud.

nutation.—Irregularities in the precessional motion of the equinoxes.

oceanography.—The application of the sciences to the phenomena of the oceans.

ocean station vessel.—A ship which remains close to an assigned position at sea to take weather observations, assist aircraft, etc.

off soundings.—In an area where the depth of water cannot be measured by an ordinary sounding lead, generally considered to be within the 100-fathom line.

overfalls.—Short, breaking waves occurring when a current passes over a shoal or other submarine obstruction or meets a contrary current or wind.

pack.—A large field of floating pieces of sea ice which have drifted together.

perigean tides.—Tides of increased range occurring when the moon is near perigee.

pilot chart.—A chart giving information on ocean currents, weather, and other items of interest to a navigator.

plot.—A drawing consisting of lines and points graphically representing certain conditions, as the progress of a craft.

plotter.—An instrument for plotting lines and measuring angles on a chart or plotting sheet.

plotting chart.—A chart designed primarily for plotting dead reckoning, lines of position from celestial observations, or radio aids, etc.

plotting sheet.—A blank chart showing only the graticule and one or more compass roses, so that the plotting sheet can be used for any longitude.

pressure ice.—Sea ice having any readily observed roughness of the surface.

primary tide station.—A place at which continuous tide observations are made over a number of years to obtain basic tidal data for the locality.

psychrometer.—An instrument consisting of suitably mounted dry-bulb and wet-bulb thermometers for determining relative humidity and dew point.

race.—A rapid current or a constricted channel in which such a current flows.

range of tide.—The difference in height between consecutive high and low tides at a place.

range of visibility.—The extreme distance at which an object or light can be seen.

Rankine temperature.—Temperature based upon a scale starting at absolute zero (−459.67 F) and using Fahrenheit degrees.

ration of rise.—The ratio of the height of tide at two places.

Reaumur temperature.—Temperature based upon a scale in which, under standard atmospheric pressure, water freezes at 0° and boils at 80° above zero.

reference station.—A place for which independent daily predictions are given in the tide or tidal current tables, from which corresponding predictions are obtained for other stations by means of differences or factors.

relative humidity.—The percentage of saturation of the air.

rise of tide.—Vertical distance from the chart datum to a high water datum, such as mean high water.

rotary current.—A tidal current which changes direction progressively through 360° during a tidal-day cycle, without coming to slack water.

sea-air temperature difference correction.—That sextant altitude correction resulting from abnormal refraction occurring when there is a difference in the temperature of the water and air at the surface.

sea anchor.—An object towed by a vessel to keep it end-on to a heavy sea or surf or to reduce the drift.

sea level.—The height of the surface of the sea.

sea mile.—Nautical mile.

seamount.—An elevation of relatively small horizontal extent rising from the bottom of the sea.

seaway.—A moderately rough sea.

secondary tide station.—A place at which tide observations are made over a short period to obtain data for a specific purpose.

seismic sea wave.—One of a series of ocean waves propagated outward from the epicenter of a submarine earthquake.

semidiurnal current.—Tidal current having two flood currents and two ebb currents each tidal day.

semidiurnal tide.—Tide having two high tides and two low tides each tidal day.

slack water.—The condition when the speed of a tidal current is zero.

solar day.—The duration of one rotation of the earth on its axis, with respect to the sun.

solar tide.—That part of the tide due solely to the tide-producing force of the sun.

solar time.—Time based upon the rotation of the earth relative to the sun.

solstitial tides.—Tides occurring near the times of the solstices, when the tropic range is especially large.

sonar.—A system of determining distance of an underwater object by measuring the interval of time between transmission of an underwater sonic or ultrasonic signal and return of its echo.

sonic depth finder.—An echo sounder operating in the audible range of signals.

sounding.—Measured or charted depth of water, or the measurement of such depth.

sounding lead (lĕd).—A lead used for determining depth of water.

sounding line.—The line attached to a sounding lead.

sounding machine.—An instrument for measuring depth of water by lowering a recording device.

sounding wire.—The wire attached to the recording device of a sounding machine.

spring range.—The mean semidiurnal range of tide when spring tides are occurring.

spring tides.—The tides occurring near the times of full moon and new moon, when the range of tide tends to increase.

stand.—The condition at high tide or low tide when there is no change in the height of water.

statute mile.—A unit of distance equal to 5,280 feet in the United States.

steam fog.—Frost smoke.

storm tide.—Increased water level due to a storm.

storm wave.—A high tide caused by wind.

stranding.—A serious grounding.

stratocumulus.—Low clouds (mean upper level below 6,500 ft.) composed of a layer of patches of globular masses or rolls.

stratus.—A low cloud (mean upper level below 6,500 ft.) in a uniform layer, resembling fog but not resting on the surface.

stream current.—A relatively narrow, deep, fast-moving ocean current.

strength of current.—The phase of a tidal current at which the speed is a maximum, or the speed at this time.

subordinate station.—A place for which tide or tidal current predictions are determined by applying a correction to the predictions of a reference station.

swell.—A relatively long wind wave, or series of waves, that have traveled a considerable distance from the generating area.

swell direction.—The direction *from* which swell is moving.

synoptic chart.—A chart showing the distribution of meteorological conditions over an area at a given time. Popularly called a "weather map".

temperature error.—That instrument error due to nonstandard temperature.

tidal current.—Current due to tidal action.

tidal current tables.—Tables listing predictions of the times and speeds of tidal currents at various places, and other pertinent information.

tidal datum.—A level of the sea, defined by some phase of the tide, from which water depths and heights of tide are reckoned.

tidal day.—The period of the daily cycle of the tides, averaging about 24^h50^m in length.

tidal difference.—The difference between the time or height of tides at a subordinate station and its reference station.

tidal wave.—The ridge of water raised by tidal action, resulting in tides at various places. The expression is popularly but incorrectly used to refer to a tsunami or storm wave which overflows the land.

tide.—The periodic rise and fall of the water surfaces of the earth due principally to the gravitational attraction of the moon and sun.

tide correction.—That altitude correction due to tilting of the surface of the sea, as by a tide wave.

tide gage.—An instrument for measuring the height of tide.

tide rips.—Small waves formed by the meeting of opposing tidal currents or by a tidal current crossing an irregular bottom.

tide station.—A place at which tide observations are made.

tide tables.—Tables listing predictions of the times and heights of tides.

tide wave.—The ridge of water raised by tidal action.

tropical cyclone.—A violent cyclone originating in the tropics.

tropic range.—The difference in height between tropic higher high water and tropic lower low water.

tropic tides.—The tides that occur when the moon is near its maximum declination, when the diurnal range tends to increase.

true wind.—Wind relative to a fixed point on the earth.

tsunami.—An ocean wave produced by a submarine earthquake, landslide, or volcanic action. Popularly called a "tidal wave" when it overflows the land.

twilight.—The periods of incomplete darkness following sunset or preceding sunrise.

undercurrent.—A current below the surface.

underwater navigation.—Navigation of a submerged vessel.

unfavorable current.—A current which decreases the speed of a vessel over the ground.

unfavorable wind.—A wind which delays the progress of a craft in a desired direction.

universal time.—Greenwich mean time.

upper air sounding.—Determination of the characteristics of the upper air.

veer.—Of the wind, (a) to change direction clockwise in the northern hemisphere and counterclockwise in the southern hemisphere, or (b) to shift aft.

velocity.—Rate of motion in a given direction.

vernal equinox.—That point of intersection of the ecliptic and the celestial equator, occupied by the sun as it changes from south to north declination, on, or about March 21, or the instant this occurs.

visibility.—The extreme horizontal distance at which prominent objects can be seen and identified by the unaided eye.

visible horizon.—That line where earth and sky appear to meet.

vulgar establishment.—The average interval of time between the transit (upper or lower) of the full or new moon and the next high water.

warm air mass.—An air mass that is warmer than surrounding air, and usually warmer than the surface over which it is moving.

warm front.—That line of discontinuity, at the earth's surface or at a horizontal plane aloft, where the forward edge of an advancing warm air mass is replacing a colder air mass.

warm sector.—An area at the earth's surface bounded by the warm and cold fronts of a cyclone.

wave.—1. An undulation or ridge on the surface of a liquid, or anything resembling this. 2. A disturbance propagated in such a manner that it may progress from point to point.

wave crest.—The highest part of a wave.

wave direction.—The direction *from* which waves are moving.

wave height.—The distance from the trough to the crest of a wave, measured perpendicular to the direction of advance.

wave length.—The distance in the direction of advance between the same phase of consecutive waves.

wave period.—The time interval between passage of successive wave crests at a fixed point.

wave train.—A group of related waves, constituting a series.

wave trough.—The lowest part of a wave, between two crests.

weather map.—Synoptic chart.

weather signal.—A visual signal displayed to indicate a weather forecast.

weather vane.—A device to indicate the direction from which the wind blows.

wind.—Moving air, especially a mass of air having a common direction of motion.

wind current.—A current created by the action of the wind.

wind direction.—The direction *from* which wind blows.

wind rose.—A diagram showing the relative frequency and sometimes the average speed of the winds blowing from different directions in a specified region.

wind vane.—A device to indicate wind direction.

wind wave.—A wave generated by friction between wind and a fluid surface.

young ice.—Newly formed ice.

Appendix B
Continuous VHF-FM
Weather Broadcasts

NOAA WEATHER RADIO

The National Weather Service provides continuous weather broadcasts to mariners within listening range of its VHF-FM radio stations in many coastal locations. These line-of-sight transmissions can be received up to 40 miles from the antenna site depending on terrain and type of receiver. Where antennas are on hills or mountains the range may extend 100 miles or more. These voice broadcasts are taped, and they repeat every 4 to 6 minutes. The tapes are up-dated periodically, usually every 2 to 3 hours, and amended as required to include the latest information.

Broadcasts include local and area warnings and forecasts, radar summaries, observations, sea conditions, and harbor water levels where available. When severe weather warnings are issued, routine transmissions are interrupted and the broadcast given over to emergency warning operations.

Following is a list of the continous weather broadcasts:

Location		*Call Sign*	*Freq. (MHz)*
ATLANTIC AND GULF OF MEXICO COASTS			
Me.	Ellsworth	KEC-93	162.40
Me.	Portland	KDO-95	162.55
Mass.	Boston	KHB-35	162.40
Mass.	Hyannisport (Cape Cod)	KEC-73	162.55
Conn.	New London	KHB-47	162.40
N.Y.	New York	KWO-35	162.55
N.J.	Atlantic City	KHB-38	162.40
D.C.	Washington	KHB-36	162.55
Md.	Baltimore	KEC-83	162.40
Md.	Salisbury (Eden)	KEC-92	162.40
Va.	Norfolk	KHB-37	162.55
N.C.	New Bern	KEC-84	162.40

	Location	*Call Sign*	*Freq. (MHz)*

ATLANTIC AND GULF OF MEXICO COASTS (cont'd.)

N.C.	Cape Hatteras	KIG-77	162.55
N.C.	Wilmington	KHB-31	162.55
N.C.	Myrtle Beach	KEC-95	162.40
S.C.	Charleston	KHB-29	162.55
Ga.	Savannah	KEC-85	162.40
Fla.	Jacksonville	KHB-39	162.55
Fla.	West Palm Beach	KEC-50	162.40
Fla.	Miami	KHB-34	162.55
Fla.	Tampa	KHB-32	162.55
Fla.	Panama City	KGG-67	162.55
Fla.	Pensacola	KEC-86	162.40
Ala.	Mobile	KEC-61	162.55
La.	New Orleans	KHB-43	162.55
La.	Lake Charles	KHB-42	162.55
Tex.	Galveston	KHB-40	162.55
Tex.	Houston	KGG-68	162.40
Tex.	Corpus Christi	KHB-41	162.55
Tex.	Brownsville	KHB-33	162.55

PACIFIC COAST AND ALASKA

Alaska	Anchorage	KEC-43	162.55
Alaska	Steward (0630–2000 Local Time)	KEC-81	162.55
Wash.	Seattle (Antenna on Green Mt. near Bremerton)	KHB-60	162.55
Oreg.	Astoria	KEC-91	162.40
Oreg.	Portland (Antenna in West Portland Hills)	KEB-97	162.55
Calif.	Eureka (Antenna on Mt. Pierce)	KEC-82	162.40
Calif.	San Francisco (Antenna on Mt. Pise)	KHB-49	162.55
Calif.	Sacramento (Antenna on Jackson Butte)	KEC-57	162.40
Calif.	Monterey (Antenna near Almaden)	KEC-49	162.40
Calif.	Los Angeles (Antenna on Mt. Wilson)	KWO-37	162.55
Calif.	San Diego (Antenna on Mt. Woodson)	KEC-62	162.40

Location		*Call Sign*	*Freq. (MHz)*

HAWAIIAN ISLANDS

	Oahu (On Mt. Kaala, 4060 ft. m.s.l.)	KBA-99	162.55
	Kauai (Kokee, 3750 ft. m.s.l.)	KBA-99	162.40
	Maui (On Mt. Haleakala, 10025 ft. m.s.l.)	KBA-99	162.40
	Hawaii (On Kulani Cone, 5500 ft. m.s.l.)	KBA-99	162.55

GREAT LAKES

N.Y.	Rochester	KHA-53	162.40
N.Y.	Buffalo	KEB-98	162.55
Pa.	Erie	KEC-58	162.40
Ohio	Cleveland	KHB-59	162.55
Ohio	Sandusky	KHB-97	162.40
Mich.	Detroit	KEC-63	162.55
Mich.	Alpena	KIG-83	162.55
Mich.	Sault Ste. Marie	KIG-74	162.55
Wisc.	Green Bay	KIG-65	162.55
Mich.	Grand Rapids	KIG-63	162.55
Wisc.	Milwaukee	KEC-60	162.40
Ill.	Chicago	KWO-39	162.55
Mich.	Marquette	KIG-66	162.55
Minn.	Duluth	KIG-64	162.55

APPENDIX C

BEAUFORT SCALE

WITH CORRESPONDING SEA STATE CODES

Beaufort number	Wind speed knots	Wind speed mph	Wind speed meters per second	Wind speed km per hour	Seaman's term	World Meteorological Organization (1964)	Effects observed at sea	Estimating wind speed — Effects observed on land
0	under 1	under 1	0.0-0.2	under 1	Calm	Calm	Sea like mirror.	Calm; smoke rises vertically.
1	1-3	1-3	0.3-1.5	1-5	Light air	Light air	Ripples with appearance of scales; no foam crests.	Smoke drift indicates wind direction; vanes do not move.
2	4-6	4-7	1.6-3.3	6-11	Light breeze	Light breeze	Small wavelets; crests of glassy appearance, not breaking.	Wind felt on face; leaves rustle; vanes begin to move.
3	7-10	8-12	3.4-5.4	12-19	Gentle breeze	Gentle breeze	Large wavelets; crests begin to break; scattered whitecaps.	Leaves, small twigs in constant motion; light flags extended.
4	11-16	13-18	5.5-7.9	20-28	Moderate breeze	Moderate breeze	Small waves, becoming longer; numerous whitecaps.	Dust, leaves, and loose paper raised up; small branches move.
5	17-21	19-24	8.0-10.7	29-38	Fresh breeze	Fresh breeze	Moderate waves, taking longer form; many whitecaps; some spray.	Small trees in leaf begin to sway.
6	22-27	25-31	10.8-13.8	39-49	Strong breeze	Strong breeze	Larger waves forming; whitecaps everywhere; more spray.	Larger branches of trees in motion; whistling heard in wires.
7	28-33	32-38	13.9-17.1	50-61	Moderate gale	Near gale	Sea heaps up; white foam from breaking waves begins to be blown in streaks.	Whole trees in motion; resistance felt in walking against wind.
8	34-40	39-46	17.2-20.7	62-74	Fresh gale	Gale	Moderately high waves of greater length; edges of crests begin to break into spindrift; foam is blown in well-marked streaks.	Twigs and small branches broken off trees; progress generally impeded.
9	41-47	47-54	20.8-24.4	75-88	Strong gale	Strong gale	High waves; sea begins to roll; dense streaks of foam; spray may reduce visibility.	Slight structural damage occurs; slate blown from roofs.
10	48-55	55-63	24.5-28.4	89-102	Whole gale	Storm	Very high waves with overhanging crests; sea takes white appearance as foam is blown in very dense streaks; rolling is heavy and visibility reduced.	Seldom experienced on land; trees broken or uprooted; considerable structural damage occurs.
11	56-63	64-72	28.5-32.6	103-117	Storm	Violent storm	Exceptionally high waves; sea covered with white foam patches; visibility still more reduced.	
12	64-71	73-82	32.7-36.9	118-133	Hurricane	Hurricane	Air filled with foam; sea completely white with driving spray; visibility greatly reduced.	Very rarely experienced on land; usually accompanied by widespread damage.
13	72-80	83-92	37.0-41.4	134-149				
14	81-89	93-103	41.5-46.1	150-166				
15	90-99	104-114	46.2-50.9	167-183				
16	100-108	115-125	51.0-56.0	184-201				
17	100-116	126-136	56.1-61.2	202-220				

Note: Since January 1, 1955, weather map symbols have been based upon wind speed in knots, at five-knot intervals, rather than upon Beaufort number.